# Setup Reduction Through Effective Workholding

# Setup Reduction Through Effective Workholding

By
Edward G. Hoffman
for the
Center for
Manufacturing Systems at
New Jersey Institute
of Technology

Foreword by
Wayne S. Chaneski

Center for Manufacturing Systems

Industrial Press Inc.
New York, New York

Library of Congress Cataloging-in-Publication Data

Hoffman, Edward G.
    Setup reduction through effective workholding / by Edward G.
Hoffman for the Center for Manufacturing Systems at New Jersey
Institute of Technology : foreword by Wayne S. Chaneski.—1st ed.
        256 p.   21.6 x 27.9 cm.
      Includes Index.
      ISBN 0-8311-3067-9
    1. Assemby-line methods. 2. Production engineering. I. New
Jersey Institute of Technology. Center for Manufacturing Systems.
II. Title.
TS178.4.H631963
658.5'33—dc20

                                                                    95-49887
                                                                        CIP

# INDUSTRIAL PRESS INC.
200 Madison Avenue
New York, New York 10016-4078

First Edition 1996

**Setup Reduction Through
Effective Workholding**

First Printing

Copyright © 1996 by Industrial Press Inc., New York, New York. Printed in the United States of America. All rights reserved. This book, or parts thereof, may not be reproduced, stored in a retrieval system, or transmitted in any form without permission of the publishers.

10 9 8 7 6 5 4 3 2

# Contents

| | |
|---|---|
| Foreword | vii |

| Part 1 | Introduction to Setup Reduction |  |
|---|---|---|
| Chapter One | Understanding Setup Reduction | 3 |
| | • Benefits of Reducing Setup Expenses | 5 |
| | • Overall Scope and Application of Setup Reduction | 5 |
| | • Analyzing the Workplace | 6 |
| | • Simplifying the Workplace Processing | 13 |
| Chapter Two | Basic Workholding Principles | 17 |
| | • General Locating Principles | 19 |
| | • Designing a Locating System | 20 |
| | • Rules for Locating | 22 |
| | • General Clamping Principles | 25 |
| | • General Construction Principles | 28 |
| | • Designing Tool Bodies | 30 |
| | • General Design Considerations | 31 |
| Chapter Three | Workholding Options and Economics | 33 |
| | • Workholding Options | 35 |
| | • Workholder Selection | 39 |
| | • Modular Fixturing Versus Dedicating Fixturing | 40 |

| Part Two | Setup Reduction Techniques | |
|---|---|---|
| Chapter Four | Techniques for Locating | 47 |
| | • Collection of hints, tips, and suggestions concerning improved methods of workpiece location | |
| Chapter Five | Techniques for Clamping | 81 |

• Collection of hints, tips, and suggestions concerning improved methods of workpiece clamping

Chapter Six   Techniques for Workholder Setups   117
• Collection of hints, tips, and suggestions concerning improved methods of workholder mounting and setup

Chpater Seven   Techniques for Chucks and Collets   145
• Collection of hints, tips, and suggestions concerning improved methods of setup involving chucks and collets

Chapter Eight   Techniques for Vises   177
• Collection of hints, tips, and suggestions concerning improved methods of setup involving vises

Chapter Nine   Techniques for Power Workholding   209
• Collection of hints, tips, and suggestions concerning improved methods of power clamping setups

Part Three   Additional Resources

Company Names and Addresses   237

Index   239

# Foreword

During my association with the Center for Manufacturing Systems at New Jersey Institute of Technology, I have met many people from a variety of manufacturing companies. These people possess a great deal of manufacturing savvy. They can look at a complicated part for a few seconds and figure out how it should be manufactured, how long it should take to make, and how much it should cost. However, these same knowledgeable people frequently underestimate the time it takes to prepare, or set up, the machines used to make these parts. More than once, I have been told that a particular setup should take "about an hour," when in practice it may take four or five times as long. As it turns out, people with the skills to machine complex parts in just a few minutes may be sacrificing many hours due to poorly planned and executed setups!

Much has been written about the machine setup process. We know we should eliminate all unnecessary or redundant activities, perform operations concurrently, move on-line operations off-line, and use the "buddy system" to minimize total setup time. All of these techniques are valuable; but they do not necessarily address the critical and time-consuming activity of preparing, or building, workholding fixtures. Through our studies at the Center for Manufacturing Systems, we found that the most labor-intensive step in the setup process is usually workholder, or fixture, preparation. This step is also cited as the one most likely to cause job delays.

It was clear from my personal observations, along with assessments performed in manufacturing companies, that help was needed on workholder preparation—specifically design, assembly, and use. With this in mind, I set out to find someone knowlegeable in these areas. It was not long before my search led me to Ed Hoffman.

I had read some of Ed Hoffman's columns in *Modern Machine Shop* and was impressed with his ideas and the way he communicated them. I decided to attend a seminar he was running entitled "Setup Reduction for Workholding." It was obvious that Ed was someone with a wealth of practical ideas about workholder design and setup time reduction. I thought he should get his ideas on paper so that others could benefit from his experience. Ed believed a book on setup time reduction, with special emphasis on workholding, could be a valuable tool; I convinced him there was a market for this tool. Hence the idea for *Setup Reduction Through Effective Workholding* was born.

This book is a collection of hints, tips, and suggestions for improving methods for workholding. These improved methods are designed to save you time and money. By using just a few of the many ideas shown, you will reduce your typical setup time and get more output from your machines.

*Setup Reduction Through Effective Workholding* is intended to be a reference book that you can keep in your shop. Leave it within reach of people responsible for setting up and running machines. Scribble in the margins, make sketches on the pages, and get it dirty! It will not only provide useful ideas, but inspire you to improve upon them.

A central theme of the book is one of Ed Hoffman's fervent beliefs: "Don't make what you can buy." Many machine shops spend a great deal of time "reinventing the wheel" because they are convinced they can build things themselves for less money. In my opinion, if you are clever enough to make something for less money than you can buy it, you should use that energy and ingenuity to improve your processing and setup techniques. Your time is better spent improving your core business. Besides, if you look at your real costs of making what you can buy, you may be surprised to find that you are not saving money after all. However, determining your real manufacturing costs is a topic better left for another time (or perhaps another book).

Wayne S. Chaneski  
Center for Manufacturing Systems  
New Jersey Institute of Technology

# PART ONE
## Introduction to Setup Reduction

# CHAPTER ONE

# Understanding Setup Reduction

# Understanding Setup Reduction

The escalating cost of manufacturing is a topic everyone talks about, but few do anything to change. Just-In-Time (JIT), for example, has wide appeal. But without proper manufacturing methods, any savings from using JIT may be offset by additional manufacturing cost. Little is gained from reduced inventory if the cost of individual setups is excessive. Although the cost of individual production runs may be reduced, the added setup costs easily drive up the total cost. Reduced setup expenses are critical to maintaining acceptable production costs.

In cost-reduction programs, workholding has not received nearly as much attention as other areas of manufacturing. Some companies view the cost of setting up jigs or fixtures as fixed values. Other companies estimate setup costs based on outdated methods and data. Seldom do companies actually study the time it takes to perform setups. So, although we spend countless hours trying to improve the cost effectiveness of the operations, we may overlook the one area—workholding—that can save the most.

In the past one-hundred years, manufacturing has made many advances in machine tools, cutting tools, and production methods; but jigs and fixtures, unfortunately, have not kept pace with these advances. Few machine tools or cutting tools resemble those from the turn of the century. Yet many of today's jigs and fixtures and workholding methods are remarkably similar to those from the 19th century. Today, as then, the main requirement of a workholder is to properly position and hold a workpiece. But, too often, today's tools and methods for workholding create major production bottlenecks. To minimize these problems, now may be an ideal time to rethink many of our old-fashioned fixturing ideas, outmoded workholding methods, and marginally effective design concepts.

## Benefits of Reducing Setup Expenses

The concept of reducing workholding setups is an idea whose time has come. Many benefits of reduced setup expenses are apparent; others are often less obvious. In addition, any benefits to a company are usually determined by the total scope of its involvement with setup reduction. The following are benefits a company can expect to achieve from reducing setup expenses.

- Lower overall production expenses.
- Reduced tooling expenses.
- Increased production speed.
- Increased production volume.
- Reduced product lead times.
- Faster production changeovers.

No matter how large or small, any company can benefit from a program of reducing setup expenses. In addition to improved overall productivity, a sound setup reduction program also makes an organization more competitive and more profitable through improvements in the complete production cycle.

## Overall Scope and Application of Setup Reduction

Setup reduction for workholding may initially seem to be merely finding ways to get parts on and off a machine tool faster. It actually involves much more. Every step, from initial product design through final inspection, must be analyzed to find better ways of producing products. Setup reduction for workholding is less a process than a state of mind. Anything to reduce the time it takes to get workholders loaded and parts into production fits this definition. The first concern in setup reduction is awareness that current methods can be improved. No matter how efficient any operation may seem, it can usually be improved by simple analysis.

Three areas of concern in the reduction of setup expenses are the *workpiece design*, the *workpiece processing*, and the *workholder design*. Although these areas are separated here, changes on one area often have significant effects in either (or both) of the other areas. For example, when a workpiece design is modified or simplified, the subsequent processing is also usually affected. When the processing is changed, the workholder design may also reflect these alterations. Although each of these areas may seem independent, in practice, a change in one area typically affects the others as well.

Improvements in any one of these areas offer many benefits. However, significant savings can be achieved by applying a systematic approach to improving all three areas. Each step, from initial product design through final inspection, must be analyzed to find better production methods. The best place to start any analysis is at the beginning, with the shop print and the manufacturing processes. Before any type of setup reduction methods can be employed, it is essential to begin from a known starting point.

## Analyzing the Workpiece

The first part of the analysis is to study the part print to see exactly what is required. Then study the operations actually performed. This might be a real eye-opening experience. In one organization, this analysis revealed four holes that were being put into a part that were not even shown on the print. Furthermore, the holes were tapped and plugged in subsequent operations in another department. Here the changes made to the machining prints were never made to the workholder, so the old hole patterns were still drilled in the parts. More astonishing, however, was that the change that removed the holes was made 12 years before.

Similarly, almost every company has had undocumented production tools and processes. Only after someone retires or leaves the company, and production is brought to a halt, does someone find the problem. Too often, emergency changes or modifications made on the shop floor are not properly recorded. Although often humorous after the fact, the results of these situations can be disastrous.

Simply studying the existing methods can often uncover significant savings. An example of this is when a company used a hit-and-miss system of storing workholders. Here the operator would get the part print and the parts delivered on a wooden pallet. The operator would then go find the workholder in a tool storage building. Since there was no consistency to the storage system, it took the operator an average of two hours to find the necessary jig or fixture. The operator would then go to one or more of the three different tool cribs to get the necessary cutters and accessories.

This added another hour to the operation time. By the time the operator returned to the machine and began the actual workholder and workpiece setup, three hours were already spent on the job. Simply changing the system to incorporate accurate records and a single centralized tool crib for all tools, workholders, cutters, and accessories reduced this time to an average of less than 30 minutes.

### Simplifying the Workpiece Design

The first step in reducing setup time and expense is designing a workpiece that can be manufactured easily. Although this may seem obvious, seldom do the methods of making parts enter into the design process. More often, a part is designed to meet only functional, decorative, or economic considerations. Although each of these is a valid concern in part design, the way the parts are made must not be overlooked during the design. To keep the cost of making the parts as low as possible, the way each design element is made should be assessed and evaluated before the element is added to the design. So, rather than making the manufacturing processes conform to the feature, each feature should be designed to meet existing manufacturing capabilities.

Ideally, product designers should have a working understanding of how the intended workpieces will be made and the capabilities of their company's manufacturing departments. The product designer is normally the best person to determine how the workpiece must be located and clamped to meet design requirements. If necessary, the designer can consult with manufacturing to determine the best way to meet both the design requirements and manufacturing capabilities.

The product designer should also question each feature or detail of a design. In this regard, the following three questions should be asked. 1) Is the feature or detail necessary or can it be eliminated? 2) Can the feature or detail be simplified? 3) Can the feature or detail be manufactured economically? Review of each design aspect not only validates and verifies the design, but will ensure that each feature can be economically produced.

## Understanding Setup Reduction

The ideal starting point for reducing workholding setups must be the workpiece being fixtured. Simplifying workpiece designs makes manufacturing easier and dramatically reduces overall tooling and production expenses. When studying a product design, the primary goal of the designer should be simplifying the product. Here, the function of the part will often determine the best design. The simple question—*what does this part do?*—sometimes leads the designer to a more efficient design.

Fig. 1-1A shows a proposed design for a cutoff blade. The part was initially designed as a one-piece unit. The unit was to be machined from 1.125" × 0.63" rectangular bar stock. The cutoff blade was hardened to RC 55, so the material specified was A2 tool steel.

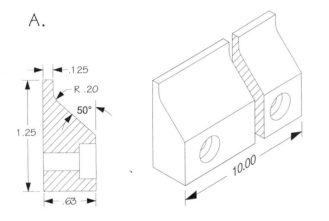

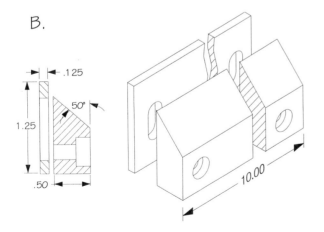

**Fig. 1-1.** Initial and modified designs of a cutoff blade.

The initial tooling estimates called for a mill fixture, a grinding fixture, a drill jig, and special form cutters. The mill fixture and special form cutters were required to machine the radius and angled area of the part. The grinding fixture dressed the top and long side of the cutting edge. The drill jig was needed to drill and counterbore six mounting holes. To reduce possible cracking in the counterbored area, a special counterbore with radiused corners was also specified.

Simple study of the function and operation of the part showed the designer that the only area of the workpiece that did any work was the 0.125" wide cutting edge. The cutting action required the blade to move less than 0.063" vertically along its long side. The remainder of the part was simply a carrier and support for the cutting edge.

Starting with this information, the designer looked for ways to change the part design to reflect the operations the part actually performed. The result was the design shown in Fig. 1-1B. Here the part is a two-piece unit. Since only the cutting edge needed to be hardened, a piece of 0.125" × 1.125", A2 precision ground stock was selected for the cutting blade. Using this material reduced both the machining and grinding requirements. The supporting bar was changed to 1018 carbon steel.

Both the slots in the blade and angle on the supporting bar were machined in vise jaw fixtures with standard end mills. Since the supporting bar was not hardened, standard drills and counterbores were used for the mounting holes in the supporting bar. The cutting blades were ground on a surface grinder using parallels and clamps.

The slotted design of the cutting blade allows the blade to be flipped over when the first side gets dull; this effectively doubled the life of the cutting blade. When the complete cutoff blade required changing, only the cutting blade was changed. The original support bar was reused. In this case, the net result of asking *what does this part do?* was a part that both worked better and cost substantially less to make. When reducing workholding setups, the starting point should always be the workpiece design. Simplifying workpiece designs makes manufacturing easier and reduces tooling and production expenses.

Another simplified design is shown in Fig. 1-2. The design called for a coupler assembly for a manual adjusting mechanism. The requirement was a rotational drive that allowed the coupler to slide along the drive shaft. The complete unit was hand driven so the rotational speed was not a concern. The original design for this coupler, as initially proposed, was a splined drive assembly as shown in part A. This design required both special spline milling cutters to machine the shaft as well as a special custom spline broach to produce the splined hole in the coupler.

After studying both the design and the specified requirements, and asking *what does this part do?* the manufacturing engineer devised a simplified alternative. The design, shown in Fig. 1-2B, is a simple hex shaft mounted in a broached hexagonal hole. This alternative accomplished the same actions as the initial design, but at a fraction of the estimated cost. Rather than using special spline cutters to machine the shaft, a standard off-the-shelf bar of 0.75" hex stock was specified. Likewise, a standard 0.75" hexagonal broach was ordered in place of the custom spline broach.

*Design Considerations*

In the design of any manufactured product, the first concerns that must be addressed are intended function and application of the item. Once these considerations are thoroughly evaluated, the next step is to determine the most efficient, cost effective, and economic design. Simplifying the manufacturing processes is an important step in reducing the overall cost of any product. The following are some general points to remember when designing a part for easier manufacturing. These general guidelines not only simplify manufacturing, but reduce the overall cost of making any product as well.

**Maximize Design Simplicity:** Every design should be made as simply as possible. Only the elements essential to the function of the workpiece should be included. Product designs should be developed for maximum simplicity in both physical and functional characteristics of the product. Workpiece or processing complexity should be reduced or eliminated at every opportunity. The complete design should be tailored to meet the specific functional requirements of the product. Simplifying the initial product design reduces workpiece complexity and, by reducing both the manufacturing and fixturing requirements, makes the product less expensive to make.

**Select Suitable Materials:** Workpiece materials should be selected on the basis of their functionality, availability, and cost. When possible, avoid exotic or specialized materials. Material selection should be based on the suitability of the material for the expected tasks. Other considerations, such as machinability ratings, heat treating operations, and

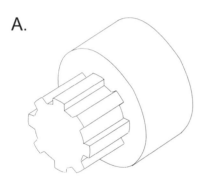

A.

ORIGINAL DESIGN

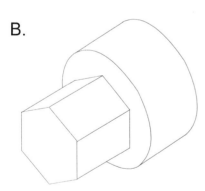

B.

SIMPLIFIED DESIGN

**Fig. 1-2.** Initial and modified designs of a coupler assembly.

general properties of the materials, must also be evaluated. As a rule, choose materials that are the least expensive to process, rather than the cheapest to buy. Some alloy steels, for example, are initially more expensive than carbon steels, but they are actually less expensive in the long run. Carbon steels often require additional processing to achieve the same characteristics of alloy steels. Hardening an alloy steel is relatively simple. Case hardening a carbon steel part could be more expensive.

**Use the Widest Possible Tolerances**: Tight tolerances only add cost to the product—not quality. Tolerances should always be relevant to the part function and requirements. They should also be within the capability of the specified processes. Any tolerance closer than 0.010" (or ±0.005") should be scrutinized to make sure that it is really needed. The general tolerances listed in the title block should be carefully analyzed to make sure the stated values are realistic. Make sure all surface texture specifications are within the capability of the process and actually necessary to the function of the part.

Typically, a simple ±0.003" tolerance on the diameter of a through hole intended for a bolt seems innocent. However, this tolerance normally requires the hole to be reamed to maintain the required limits of size. In most cases, bolt holes do not require this degree of accuracy. Likewise, a ±0.010" tolerance specification on the overall length of a shaft when the length is not critical is unnecessary. This simple specification requires a secondary finishing operation that adds to the cost of the processing, with no functional benefit. A better approach would be to specify a larger tolerance and simply deburr the ends.

When tolerances are applied to workholders, there is a wide difference in the amount and application of tolerances. Some companies assign an arbitrary tolerance value, such as ±0.0002", on all tool dimensions. Others simply apply a percentage tolerance, like 10%, to all print tolerances.

Assigning arbitrary tolerance values to jigs or fixtures is a common practice. But it does not express the true relationship of the tooling tolerances to the part dimensional tolerances. If, for example, one part dimension has a tolerance of ±0.0003" and an arbitrary tolerance of ±0.0002" is applied, the resulting tool tolerance is 67% of the part tolerance. Conversely, if another dimension has a tolerance of ±0.020", the resulting tolerance would be 1% of the part tolerance. This wide variation could cause some of the tooling dimensions to be too accurate and others not accurate enough. Arbitrary tolerances, while valuable, many times cannot reflect the true design intent for tooling applications.

Percentage tolerances, unlike arbitrary tolerances, can be used to accurately reflect the relationship between the workpiece tolerances and the tooling tolerances. When tooling tolerances are specified as a percentage of the part tolerances, the true relationship between the tool and the part is easily maintained. But, here again, if percentage tolerances are not properly used and applied, they too can drive up the cost of a special tool.

Many companies who use percentage-type tolerances for their tooling apply a base percentage of 10%. This percentage value means the tolerance applied to the tool is equal to 10% of the part tolerance for the feature. So, a ±0.010" part tolerance results in a tooling tolerance of ±0.001". While acceptable, a 10% tolerance is normally associated with gauges rather than workholders. A more appropriate tolerance for jigs and fixtures would be in the 50% range. A 50% tooling tolerance permits lower cost tooling while maintaining an adequate accuracy. This is especially true of limited production tools which are used for lots of less than 50 to 100 parts. After all, if a print dimension has a tolerance of ±0.010", and the finished part is within ±0.005" is it any less acceptable than a part within ±0.001"?

**Standardize Where Possible**: Standardize features, methods, and hardware wherever practical. To simplify designs and reduce costs, choose standard components instead of specialized components. Reduce the inventory of common hardware items such as screws, bolts, and washers by deciding on a limited number of standard sizes and styles of each fastener. Likewise, material selections should also be limited to a range of standard types, sizes, and shapes.

Standard size cutters and tooling should always be used in place of specialty items. Standard cutters are much less expensive than specials. Limit the tooling inventory by deciding on standard tap drill sizes, body diameter drills, and counterbores for each thread size. One standard set of cutting tool sizes for each thread size allows larger quantities of the same tools to be purchased, thus reducing the per tool cost.

Common and repetitive machining operations that do not have a direct bearing on part function should also be standardized. Processes such as chamfering should have standard drawing specifications; this reduces problems encountered with unnecessarily tight chamfer specifications. Other machining specifications that should be standardized are:

- undercuts, grooves, O-ring seats, and thread runouts;
- radii, fillets, and rounds;
- countersinking, counterboring, and spotfacing operations;
- decorative or press-fit knurling operations.

**Reduce Production Steps and Minimize Handling**: Where possible, workpieces should be designed to do as many operations as practical from each setup. When the part design permits, gang operations together. Design workpieces so individual operations can be minimized. If volume allows, and standard cutters are unsuitable, special cutters may be used to combine machined details. Where acceptable, try to combine secondary machining or finishing operations.

Rather than sawing and finishing a cut surface in a secondary operation, milling cutters may sometimes be used for cutoff operations. This combines both the cutoff and finishing into a single process, and eliminates the requirement for the secondary machining operation. Special machining or finishing operations on unimportant areas of the workpiece should also be eliminated. The following are a few ideas that should be considered to reduce the production steps and parts handling.

- Design parts for bar stock where possible. During machining, the bar holds the part.
- If practical, mount parts on fixturing plates.
- Consider using palletized arrangements.
- When possible, use automatic feeding devices.

Always balance the precision and production needed against the cost and capability of the production methods. Ensure that the production methods used are compatible with the specified requirements.

### Design Characteristics to Avoid

In addition to following sound workpiece design principles, there are a few design characteristics you should avoid, if possible.

**Unusual or Overly Complex Shapes**: Workpiece designs that contain features or details that are either odd shaped or extremely detailed are also very expensive to produce. When possible, simplify these features to make manufacturing easier and less expensive. Although a feature may appear quite simple on paper, when applied to the actual workpiece, the detail may tend to be considerably more complicated.

Casting and forging are common processes used to make parts. Although widely used, cast and forged parts are typically quite difficult to fixture; this is due to the lack of uniform and consistent locating and clamping surfaces. But, with just a little foresight during the initial design, even these parts can be made more manufacturable.

## Understanding Setup Reduction

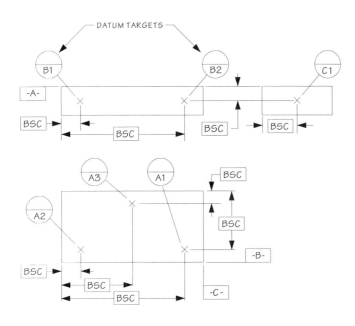

**Fig. 1-3.** Datum targets specify the exact position of each workpiece locator.

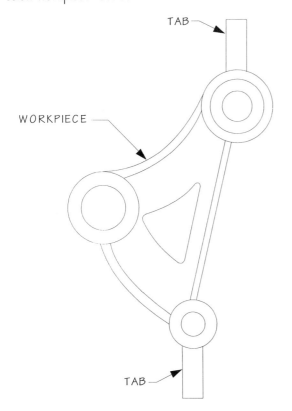

**Fig. 1-4.** Tabs cast or forged directly on the workpiece are a method of simplifying the location of complex cast or forged workpieces.

The first step in designing cast or forged parts that are easier to make is specifying the locating points. Simply specifying how the part should be located eliminates many of the assumptions and guesses that tool designers are often called on to make. Making these locating decisions during the initial part design also permits the designer to find and fix any design problems that might affect the way a part is fixtured.

The methods used to specify part location are determined both by the part itself and the operations performed on the part. The simplest method of specifying part location is by using datum targets (Fig. 1-3) to indicate how the location is accomplished. Datum targets indicate the exact position of each locator. An enlarged boss may also be specified to assure adequate space for the location. Incorporating a cast or forged boss into thinner workpieces, where a hole is to be used as a locator, can also guarantee sufficient thickness for proper location.

Cast tabs are another method that may be used to simplify the location of complex cast parts. The tabs are cast or forged directly on the workpiece, as shown in Fig. 1-4. These tabs are then machined to suit the locational methods, and are used for locating and clamping the workpiece throughout the machining. Once the workpiece is complete, the tabs are removed. This method works well for parts that are not expected to require remachining during their service life.

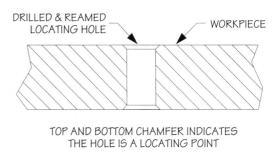

**Fig. 1-5.** Special tooling holes may also be used to locate the workpiece. These holes may be identified with a shallow counterbore at the top and bottom of each hole.

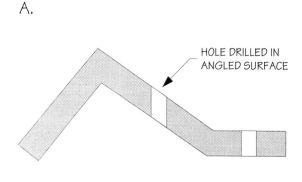

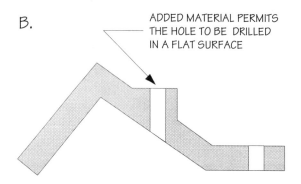

**Fig. 1-6.** Simply adding more material to a casting can simplify drilling holes in angled surfaces.

For workpieces that may need remachining, or where the part design permits, special tooling holes may also be used in place of the tabs to locate the workpiece (Fig. 1-5). The locating holes are drilled and reamed in nonfunctional areas of the workpiece. These holes then locate the workpiece for either initial machining or remachining. The tooling holes should be designated on the drawing and on the workpiece to clearly identify them as locating points. One method of identifying these locating holes in the workpiece is with a shallow counterbore at the top and bottom of each hole.

**Angled or Skewed Holes and Features:** Holes are quite possibly the most common type of machined detail. They are found in almost every conceivable area of a workpiece. But placing these holes in areas that are difficult to machine can often add significantly to the total cost of the workpiece. Fig. 1-6A shows a workpiece where a hole placed on an angled surface caused problems. Here the area was first milled to provide a flat for the hole. This added step more than doubled the cost of producing the feature. By simply adding a bit more material to the casting, as shown in Fig. 1-6B, the hole was drilled in a single step, thus reducing the cost. The cost of adding the additional material during the casting process added just a few cents to the total cost of the casting.

**Inaccessible Features or Datums:** One recurring problem for tool designers is locating the workpiece. This is particularly true when workpieces are referenced to inaccessible or imaginary datums. These datums usually take one of two forms: locating points at inaccessible points on the workpiece, or center lines in holes or centrally located part axes. Imaginary datums are great on paper, but it is very difficult to make a physical reference to a datum that is not really there.

The solution to this problem is to only use physical datums. By specifying the appropriate surfaces relative to center lines, and making the necessary allowances for slight deviations, parts will not only be easier to fixture but the possibility of high reject rates is greatly reduced.

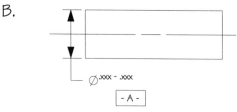

**Fig. 1-7.** To avoid locating problems, only use physical features as datums. Instead of specifying a center line as a datum, it is often better to call out the periphery of the hole as the datum.

# Understanding Setup Reduction

Generally, when a referenced datum is specified on a nonexistent center line, as shown in Fig. 1-7A, the periphery of the feature is used for location. However, due to many factors, the periphery may sometimes not be an appropriate locating, or reference, point without further control or clarification. So, rather than making the center line the datum, it is often better to call out the periphery of the hole as the reference (Fig. 1-7B). Then apply the necessary controls to this periphery to ensure that the related features are positioned as desired.

Simply checking to make sure that all datum specifications are reasonable and within manufacturing capabilities both simplifies the necessary workholders and often substantially reduces overall production costs.

## Simplifying the Workpiece Processing

The workpiece design will usually determine the specific processes performed on a workpiece. Many times, problems with a particular design will not appear until the processing operations are specified. The workpiece processing is usually decided by a Process Engineer; however, everyone involved with the workpiece is also responsible to make sure the machines and processes specified are the best available. If an

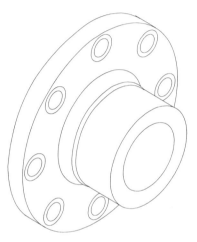

THE MOUNTING HOLES WERE COUNTERBORED IN THE ORIGINAL WORKPIECE DESIGN

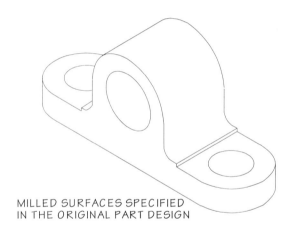

MILLED SURFACES SPECIFIED IN THE ORIGINAL PART DESIGN

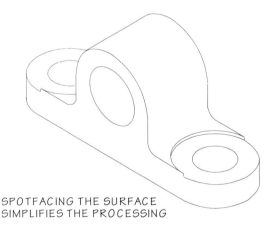

SPOTFACING THE SURFACE SIMPLIFIES THE PROCESSING

**Fig. 1-8.** Changing a milling operation to a spotfacing step substantially reduced the processing costs.

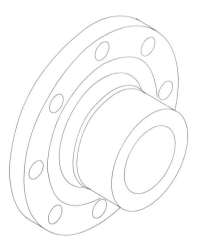

FACING THE MOUNTING SURFACE SIMPLIFIES THE PROCESSING

**Fig. 1-9.** Counterbored holes were eliminated by simply facing the complete flange.

alternate method of performing a task would be better, it should be suggested. There is no one correct way to do any task. The ideal condition is when the methods and cost can be balanced to achieve the best processing at the least expense.

In simplifying the processing of any workpiece, altering simple details can often have a dramatic effect on the total cost of producing the product. Fig. 1-8 shows how simply changing a milling operation to spotfacing substantially reduced the processing costs. Here the original specification called for a milled surface on the mounting flange of the cast bracket. The only reason for this surface was to produce a flat seat for the fastener. Changing the specification to a spotfaced area around the holes produced the same flat mounting surface at a much reduced cost. Here both the drilling and spotfacing are performed on the same machine with only minimal changeover time. The older method required a completely different setup on another machine tool.

Another similar example of altering the processing to eliminate unnecessary operations is shown in Fig. 1-9. In this case, the original specification called for the holes to be counterbored in the cast flange. By studying the processing, the designer determined that simply adding a facing step to the turning operation performed on the workpiece before the drilling not only eliminated the need to counterbore the holes, but also made the drilling operation easier.

In some cases, the basic workpiece design should be altered to simplify the processing operations. The workpiece shown in Fig. 1-10 has a ground diameter and a ground shoulder. In the initial design, the shoulder had a large radius that required a specially shaped grinding wheel to grind both the shoulder and the radius. Since there was so much surface contact, the grinding wheel required frequent dressing. Simply replacing the radius with an undercut achieved the same result at a

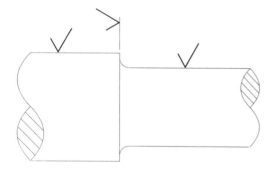

THE RADIUS MAKES THE WORKPIECE VERY DIFFICULT AND EXPENSIVE TO GRIND

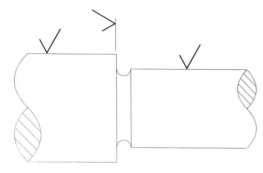

THE UNDERCUT SIMPLIFIES THE GRINDING OPERATIONS

**Fig. 1-10.** Occasionally, the basic workpiece design should be altered to simplify the processing operations.

A.

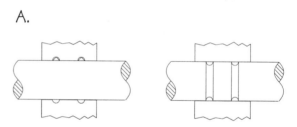

MACHINING THE "O"-RING GROOVES IN THE SHAFT INSTEAD OF THE BORE SIMPLIFIES THE PROCESSING

B.

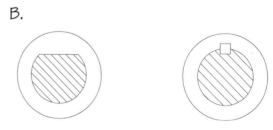

REPLACING THE D-SHAPED HOLE WITH A KEYED ASSEMBLY SIMPLIFIES THE PROCESSING

**Fig. 1-11.** Simple changes to the workpiece design can substantially reduce the processing expenses.

fraction of the cost. Here a standard grinding wheel was used and dressed much less frequently.

Fig. 1-11 illustrates two other processing alterations that greatly reduced the overall cost of the operations. The arrangement shown in Fig. 1-11A is a shaft assembled in a bore. The original design specified two O-ring grooves to be bored into the hole. This operation was quite time consuming and expensive. Since the reason for the O-rings was just to prevent leakage through the assembly, the location of these grooves is not a critical factor. In the new design of this assembly, the O-ring grooves were machined in the shaft. This design change simplifies machining the grooves, and significantly reduces the time required to produce these details.

The assembly shown in Fig. 1-11B was designed to prevent rotation of the shaft inside the bored hole. The "D"-shaped hole of the original design required a special broach to produce the hole. The revised design uses a keyed arrangement to prevent the rotation. This arrangement substantially reduced the overall cost of the unit, and no special cutters or other equipment were required.

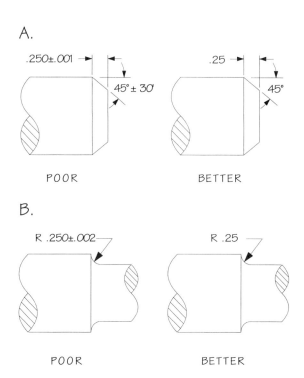

**Fig. 1-13.** Unnecessarliy tight tolerance can also have an effect on the workpiece processing.

When chamfers and radii are specified on a workpiece, every effort should be made to standardize these details. As shown in Fig. 1-12A, two different size chamfers are specified for this shaft. Since the only purpose of the chamfer at the end of the workpiece is to break the sharp edge, simply changing the specification from a 45° chamfer to a 35° chamfer eliminates the need for a second tooling setup. Here the tool used for the 35° chamfer is simply moved over to break the corner.

This same procedure can be applied to radii, as shown in Fig. 1-12B. Here the 0.15" radius is the more important one, and the 0.13" radius is intended only to provide a transition between the two diameters. Changing the 0.13" radius to a 0.15" radius has no effect on the function of the workpiece, but it eliminates a tool change.

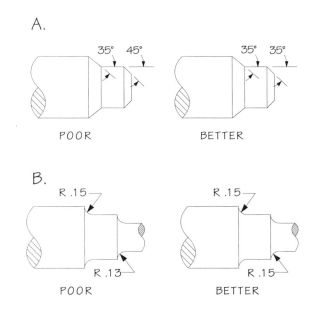

**Fig. 1-12.** Standardizing details can both simplify the processing and reduce the overall cost.

Unnecessarily tight tolerance can also affect the processing. As shown in Fig. 1-13, both the chamfer shown in part A and the radius in part B are specified with tight tolerances. Where possible, these tolerances should be opened up to permit the features to be machined more quickly and easily. If there is no critical function for these features, the tolerance should be as wide as possible.

When machining internal cavities or openings, one technique that will both simplify the processing and cut the machining cost is specifying the correct corner radii. The corner radii specified for these features, where practical, should match the cutter expected to be used to machine this feature (Fig. 1-14). This eliminates the need to change cutters just to machine a set of radii that have no effect on the function of the workpiece.

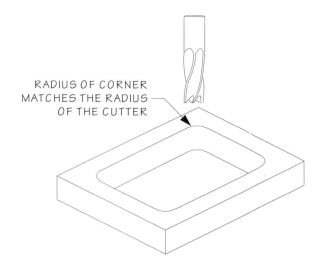

**Fig. 1-14.** To eliminate unnecessary cutter changes, specify corner radii sizes that match the cutter used to mill the cavity.

# CHAPTER TWO

# Basic Workholding Principles

# Basic Workholding Principles

Following the established principles of good design is the most important consideration in designing and building any workholding device. Few jigs or fixtures will function as required or intended without first being properly constructed. Although there are many variations of workholders, each must conform to a few basic design principles to guarantee proper operation and overall safety. Locating and clamping are two general areas of workholder design that should be completely understood to assure successful workholder designs.

## General Locating Principles

Locating is a primary function of any workholder. To properly design the specific locating elements, both the overall function and purpose of locating should be completely understood. To perform properly, a workholder must ensure the repeatability of the operation. Repeatability is the ability of a workholder to properly position the part so it can be processed, part after part, within the stated dimensional limits.

To guarantee the repeatability, locators must perform the three primary functions of workholding—they must hold, support, and locate the workpiece. When these three functions are achieved, the repeatability of the process is assured. To identify the functions of location and the importance of the locators in workholding, these three functional responsibilities of locators must be completely understood.

### Holding the Workpiece

Holding the workpiece, although often thought of as a function of clamping, is actually the responsibility of the locators. The major function of locators is to properly position the workpiece. Once positioned, however, the locators must also maintain the workpiece location and resist the cutting forces. To do this, the locators must prevent any movement of the part caused by the cutting tool acting on the workpiece.

The only real function of the clamps in a workholder is to hold the workpiece against the locators. Locators, when properly applied, provide the workpiece with a positive stop to resist any movement. Clamping devices hold the part with only the friction between the clamp and the workpiece. So, in practice, the part is held in place with the locators and held against the locators with the clamps.

### Supporting the Workpiece

A workpiece support is a locator that most often carries the weight of a workpiece. In common fixturing setups, the supporting elements are positioned to locate the workpiece from below. Although the term *support* is applied to these elements, they are still a form of locator. Only the relative position of the locating elements to the workpiece position determines whether the elements are identified as supports or locators.

In most fixturing arrangements, the supports reference the primary locating feature, or surface, of the workpiece. As shown in Fig. 2-1, this surface is typically the bottom reference surface of the workpiece. Here the supports reference the workpiece position in the Z-axis.

The different workpiece axes are customarily identified by a series of letters (the most common identifications are X, Y, and Z). The X-axis of the part describes the longitudinal horizontal axis; the Y-axis identifies the traverse horizontal axis; and the Z-axis indicates the vertical axis of the workpiece.

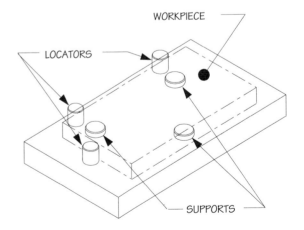

**Fig. 2-1.** Locators and supports.

*Locating the Workpiece*

Locating is simply the process of properly positioning the workpiece in the workholder with relation to its horizontal axes. Where "supporting" describes the vertical position of the workpiece, "locating" refers to the horizontal position of the workpiece. Supporting references the workpiece along the Z-axis, and locating references the workpiece relative to the X- and Y-axes. Although two terms—supporting and locating—describe these types of workpiece referencing, both may also be referred to as locating.

## Designing a Locating System

To accomplish the three functional responsibilities of location, the locators must be properly selected and positioned. Locators must be correctly positioned with respect to the *degrees of freedom of the workpiece*. Understanding and properly applying the *degrees of freedom* concept is very important in achieving the most effective and accurate workpiece location.

Any workpiece, free in space, is capable of moving in an unlimited number of directions. To limit workpiece movements to a more manageable range and to simplify location, only twelve possible movements are considered in fixture design. These movements are referred to as the twelve degrees of freedom. As shown in Fig. 2-2, the twelve degrees of freedom are directly related to the axes of the workpiece. Since there are three workpiece axes (X, Y, and Z) and twelve degrees of freedom, each axis has four degrees of freedom, as shown. On each axis, two degrees of freedom are axial and two are radial. Axial degrees of freedom permit lateral movement along each axis. The radial degrees of freedom permit rotational movement around each axis.

*The Six-Point Locating Method*

To properly fixture a workpiece, all twelve degrees of freedom must be restricted. The simplest and most efficient method to restrict most workpiece designs is with the six-point, or 3-2-1, locational method. These six locating points restrict a total of nine degrees of freedom. The remaining three degrees of freedom can be restricted with a clamping device.

The six-point locational method is best for workpieces having a rectangular form (Fig. 2-3). These workpieces are located by their outer surfaces. The primary workpiece surface is located on three points, as shown in part A. Three points are used here since three points establish a plane. Once these points are located, five degrees of freedom are restricted. As shown, the workpiece cannot move down, nor can it rotate radially about either the X- or Y-axes.

The secondary locating surface is perpendicular to the primary locating surface. This surface is located on two points, as shown in part B. These two locators further restrict an additional three degrees of freedom, for a total of eight. As shown, these locators restrict the axial workpiece movement in one direction on the Y-axis and the radial movement around the Z-axis.

The tertiary, or third locating surface, is perpendicular to both the primary and secondary locating surfaces. Here, as shown in part C, a single locator restricts a single degree of freedom. This locator restricts the axial movement in one direction along the X-axis. Together, these six locators restrict a total of nine degrees of freedom. The remaining three degrees of freedom are contained with a clamping device.

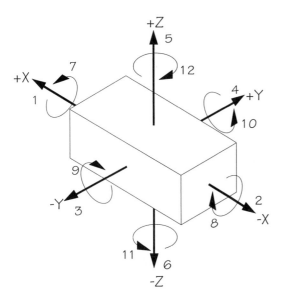

**Fig. 2-2.** The twelve degrees of freedom.

# Basic Workholding Principles

## Alternate Locating Methods

Any of several methods can be used to locate a workpiece. However, finding the most effective and efficient location should be the goal in designing any locating system. The workpiece usually determines the locational method employed in design of any jig or fixture. However, some situations permit a choice of locating methods. Here, understanding the preference of locating features is the first step in selecting a suitable locating method.

Although the six-point locating method described above is the most common method for locating workpieces, it is not always the most efficient method. With the six-point method, six locators are needed to restrict nine degrees of freedom. However, if a workpiece is located by a single hole (Fig. 2-4), one locating pin can restrict nine degrees of freedom.

Using the centrally mounted primary locator, as shown, the workpiece cannot move axially, downward along the Z-axis, nor can it move axially or radially along or about either the X- or Y-axes. The only directions this workpiece can move are radially about the Z-axis and axially, upward along the Z-axis.

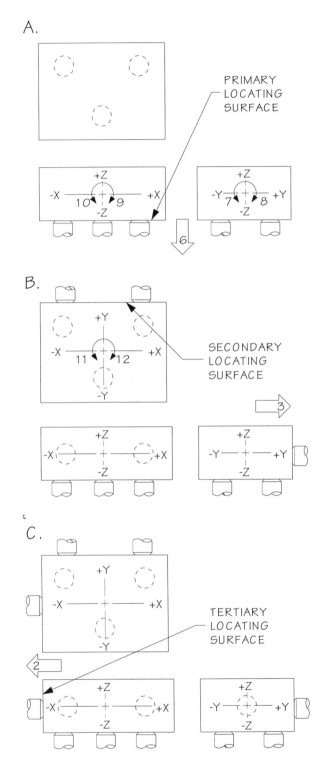

**Fig. 2-3.** The six-point, or 3-2-1, locational method.

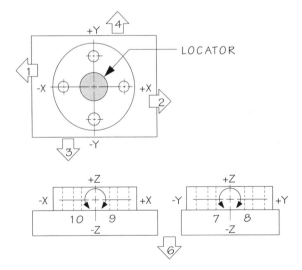

**Fig. 2-4.** A single locating pin, mounted in a hole, can restrict nine degrees of freedom.

If the workpiece had a second hole, as shown in Fig. 2-5, the hole may act as a radial locator. When both the center hole and second hole locate the workpiece, a total of eleven degrees of freedom are restricted. As before, the primary locator restricts nine degrees of freedom. The second locating pin restricts radial movement of the workpiece, in both directions, about the Z-axis. The only direction this workpiece is free to move is axially, upward along the Z-axis. As before, this movement is restricted with a clamping device.

### Rules for Locating

The overall position, orientation, and relationship of the locators is very important to the effectiveness and accuracy of any workholder. The locators must be carefully planned into the design and construction of the workholder. The following are some of the design considerations to remember and problem areas to avoid when designing locators.

*Selecting and Positioning Locators*

When choosing the locators for a workpiece, the locators must be sized and positioned to suit the specific requirements of the workpiece. The locator must be large enough to provide ample workpiece support and sufficient rigidity to resist all clamping and tooling forces. Likewise, the locator size must always be based on the Maximum Material Condition (MMC) size of the locating feature. When locating on an external feature, such as a diameter, the MMC size is the largest size of the feature allowed by the specified dimensions. If the locating element is an internal feature, such as a bored hole, the MMC size is the smallest size of the hole allowed by the specified dimensions.

When possible, the locators should always contact the workpiece on a machined surface. This helps to ensure the necessary locational repeatability from one workpiece to the next. Likewise, the locators should also be positioned away from dirt and chips. When this is not possible, slots or holes should be provided in the workholder to permit the dirt, chips, and coolant to be easily removed so they do not interfere with the workpiece location.

Another point to remember when positioning locators and supports is to place the elements as far apart as practical; this provides better locational stability and helps to minimize workpiece movement. This is particularly important when positioning supports for large workpieces. Here the locators, or supports, under the workpiece should be positioned to provide maximum stability for the part. When practical, the supports should be positioned to permit the operations to be performed within the area inside the supports (Fig. 2-6). For applications where the shape of the workpiece inhibits the placement, auxiliary supports help prevent workpiece distortion.

*Locating Tolerances*

Manufacturing and tooling tolerances are a primary concern in every aspect of production. Throughout industry there is a wide disparity in the amount and application of tolerances assigned to workholders. Some companies assign an arbitrary tolerance value, such as ±0.0002", on all tool dimensions. Other companies simply apply a percentage tolerance, such as 10%, to all print tolerances. The specific form and amount of tolerance applied to a workholder is most often determined by the workpiece. However, to ensure the necessary accuracy, the following suggestions should be used as a guide in applying locational tolerances to any workholder.

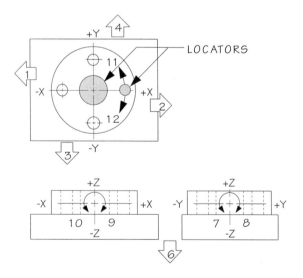

**Fig. 2-5.** Two hole mounted locating pins can restrict eleven degrees of freedom.

Assigning arbitrary locational tolerance values, while a common practice, does not express the true relationship of the tooling tolerances to the workpiece dimensional tolerances. If, for example, one part dimension has a tolerance of ±0.0003", and an arbitrary tolerance of ±0.0002" is applied, the resulting tool tolerance is 67% of the part tolerance. Conversely, if another dimension has a tolerance of ±0.020", the resulting tolerance is 1% of the part tolerance. This wide variation results in some tolerances that are too accurate and others that may not be nearly accurate enough. Arbitrary (or fixed) tolerance values, although useful for some manufacturing applications, cannot reflect the true design intent or relationships for workholder designs.

Percentage tolerances, unlike arbitrary tolerances, can accurately reflect the relationship between the workpiece tolerances and the workholder tolerances. Specifying tooling tolerances as a percentage of the part tolerances clearly shows the true relationship between the workholder and the workpiece. But, here again, if these percentage tolerances are not properly applied, they too can drive up the cost of a workholder.

Some companies that use a percentage tolerance for their fixturing apply a base percentage of 10%. This percentage value means the tolerance applied to the workholder is equal to 10% of the tolerance for the workpiece feature. So, a ±0.010" workpiece tolerance results in a workholder tolerance of ±0.001". While acceptable, a 10% tolerance is normally associated with gauges rather than workholders. A more appropriate, and much less expensive, tolerance for jigs and fixtures is in the range of 30% to 50% of the workpiece tolerance.

A tooling tolerance in the 30% to 50% range permits lower fixturing cost while maintaining adequate workpiece accuracy. This is especially true of limited production tools employed for lot sizes of less than 50 to 100 parts. After all, if a print dimension has a tolerance of ±0.010", and the finished part is within ±0.005", it is just as acceptable as a part within ±0.001". However, the specific tolerances applied to any workholder must always be determined by the parameters of the workpiece. In cases where closer tolerances are necessary, they should be specified. But, where possible, to reduce the cost of the workholder, the tolerances should be opened up as much as practical.

### Duplicate Location

Duplicate, or redundant, location takes place when a workpiece is located on more than one related surface or feature. As shown in Fig. 2-7, this can occur with parallel features, concentric features, or other similar arrangements of related features. The main problem with duplicate location lies in trying to locate a workpiece on multiple locating features. This creates problems in accurately and repeatably locating a workpiece. Also, workpieces mounted on multiple locating features are often difficult to load and unload in the workholder.

The workpiece dimensions normally indicate the locating or datum surfaces. When datum information is not directly shown, the surfaces for dimensioning a feature may also be used for locating the workpiece (Fig. 2-8). Many times the function of the workpiece helps decide how it is located. When designing a locating system, the first step is to identify the proper reference surface or feature for each locational plane or

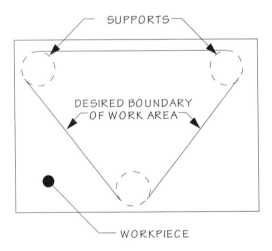

**Fig. 2-6.** Supports should be positioned to allow all operations to be performed within the area inside supports.

axis. Once this feature is identified, only this feature should be used to locate the workpiece.

## Foolproofing the Location

One of the principle responsibilities in designing any jig or fixture is to make sure the workpiece is properly loaded into the workholder. Workpieces are often improperly machined simply because they were incorrectly loaded in a workholder. Any workholder that allows workpieces to be loaded incorrectly can present serious problems in the production department—not to mention the number of scrapped workpieces. The proper way to load the part into the workholder may seem obvious to the designer or an experienced machine operator. Too often, however, the person actually doing the work may not apply the same logic to the task.

Do not assume that the operator knows what to do.

Design the workholder to make sure the workpiece is properly loaded into the workholder. Ensuring that the workpiece can be loaded only one way is commonly referred to as *foolproofing the workholder*. There are any number of ways that can be used to foolproof a workholder, but the simplest way is almost always the best option.

Before designing a foolproofing device, the need for such a device must first be established. Not every workpiece design requires a foolproofing device. As shown in Fig. 2-9, if a workpiece has identical features, it may not matter which way the workpiece is loaded. If, however, the workpiece can be positioned improperly, a foolproofing device may be required. The workpiece shown in Fig. 2-10 does have a specific way it should be mounted. Here a simple pin is installed to prevent the workpiece from being loaded if it is flipped end to end or

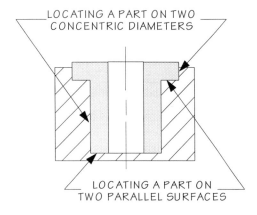

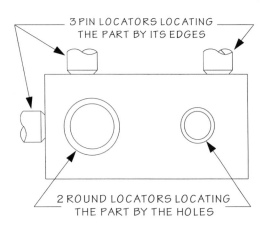

**Fig. 2-7.** Duplicate, or redundant, location.

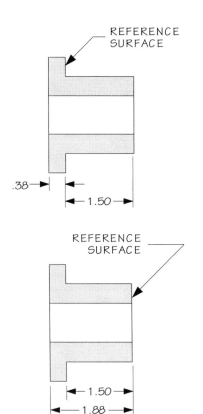

**Fig. 2-8.** The dimensioned surfaces may be used for locating the workpiece when datum information is not directly shown.

# Basic Workholding Principles

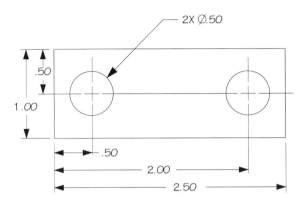

**Fig. 2-9.** Not every workpiece design requires a foolproofing device.

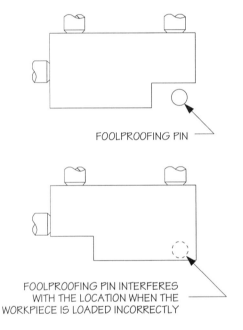

**Fig. 2-10.** A foolproofing device prevents incorrectly loading the workpiece.

top to bottom. Often a device as simple as a pin is all that is needed.

In some cases, the locators themselves may serve to both position the workpiece and foolproof the location. For example, locating a workpiece relative to a center axis is quite easy. As shown in Fig. 2-11A, the center hole may be used as a concentric locator. The two dowel pins act as a radial locator to prevent rotation about the center axis. In this arrangement, the pins used for radial location of the workpiece also act as a foolproofing device to prevent improper loading. If the workpiece has a second hole (Fig. 2-11B), the center hole may be used as a concentric locator, and a second hole serves as a radial locator to prevent rotation about the center axis.

### General Clamping Principles

Securely holding the workpiece is an essential function of any workholder. So understanding the function of clamping is an important part of design. The first function of a clamp is to securely hold the workpiece against the locators. The second (and equally important) function is to resist all secondary tool forces generated in the operation.

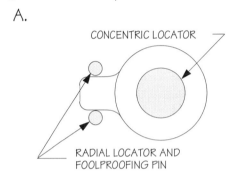

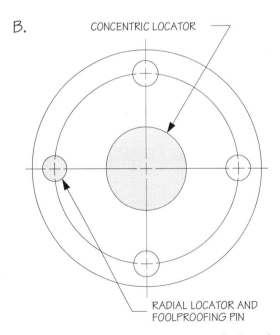

**Fig. 2-11.** The locators may also serve to foolproof the location.

Although clamps are often thought to hold the workpiece in a jig or fixture, this is not their proper application. Holding the part, as mentioned earlier, is the responsibility of the locators. Here, the locators hold the position of the workpiece and resist the primary cutting forces. To do this, the locators should prevent any movement of the part caused by the forces of the cutting tool on the workpiece. The only function the clamps should serve is to hold and maintain the position of the workpiece against the locators. In a properly designed and constructed workholder, the locators and clamps work together to correctly position and hold the workpiece.

### Tool Forces

The tool forces generated by an operation must be resisted by both the locators and the clamps. Tool forces may be divided into *primary tool forces* and *secondary tool forces*. Primary tool forces are resisted by the locators and secondary tool forces are resisted by the clamps. Fig. 2-12 shows the difference between these two types of tool forces.

The tool forces that the clamps must resist are the secondary, rather than the primary, tool forces. In the drilling operation shown here, for example, the primary tool forces are generated in both a downward and radial direction. The locators for this workholder should be large enough to resist these primary tool forces. The secondary tool forces act against the workpiece as the drill breaks through the workpiece. At the point when the drill breaks through the workpiece, the natural tendency is for the workpiece to lift up (or climb) the drill. The clamps need to resist only these secondary forces.

### Clamping Forces

Clamping forces are the forces generated by the clamps against the workpiece during the clamping operation. In some cases, if the workpiece is not properly supported, clamping forces can distort or deform the workpiece. When this occurs, the damage is usually discovered only after the workpiece is unclamped and inspected.

To minimize the negative effects of the clamps, the clamps should always be positioned over a naturally rigid or supported area of the workpiece. The areas selected to support the workpiece should allow the workpiece to be referenced and clamped without any chance of distortion once the workpiece is clamped.

The ideal position for a clamp is directly over the supports (Fig. 2-13)—at no time should a clamp be positioned between two supports. The leverage of the clamp acting against the supports could very easily deform the workpiece. In situations where the workpiece

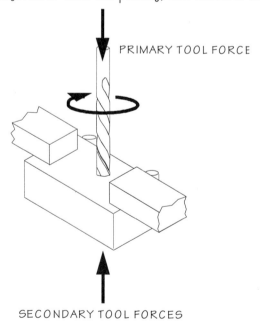

**Fig. 2-12.** Primary tool forces are resisted by the locators. Secondary tool forces are resisted by the clamps.

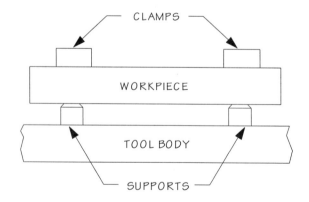

**Fig. 2-13.** Clamps should be positioned directly over a support.

# Basic Workholding Principles

is mounted directly to the tool body, without a separate support, the clamp should be positioned directly over the thickest area of the part. Here, the thickness of the workpiece usually provides a naturally rigid area and reduces any possibility of distortion.

Unfortunately, clamping directly over the thickest area of the workpiece is not always possible. In these cases, an auxiliary support is required to prevent damage to the workpiece. Once again, the specific design of the workpiece determines how the support is designed. Fig. 2-14 shows a few supports that may be applied to various workpiece designs.

### Selecting and Positioning Clamps

Clamps must be properly positioned and correctly applied to maintain the positional accuracy of the complete setup. The clamping operation, like locating and supporting, requires an understanding of a few basic principles.

Safety and clamping efficiency are primary concerns in selecting clamps. All clamping devices must be completely safe and reliable. Likewise, the clamps selected must also be as efficient as possible. The safety of the operator is a major factor in choosing clamps. Some clamp styles work very well for some applications and not so well for others. When making a selection, evaluate both the safety and efficiency of the clamping operation.

Screw clamps, for example, are a very safe form of clamp. Screw clamps, however, are not nearly as efficient as other forms of clamps. This inefficiency is due to the basic action of a screw—the mechanical action of a screw is not as fast or efficient as a lever or other mechanical action. However, a screw provides a very secure holding action.

Cam clamps, on the other hand, are a very fast and efficient clamp; however, cam clamps may not be as

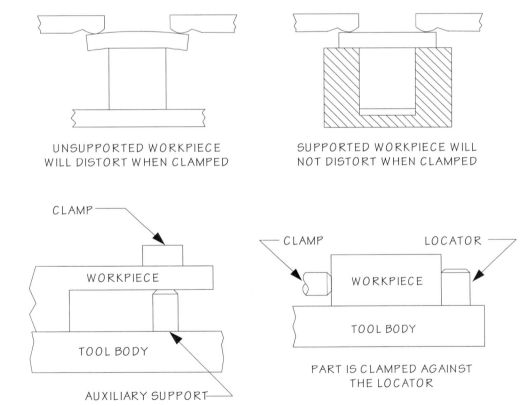

**Fig. 2-14.** The workpiece usually determines how the support is applied.

safe as other clamp designs. Cam clamps rely on the friction between the clamp face and the workpiece to provide the holding action. Unfortunately, the security of the holding action can easily be affected by excessive vibration. In situations where considerable vibration is expected, such as with milling operations, other kinds of clamps may be better suited.

Positioning the clamps is another concern in designing a workholder. The clamps should always be positioned so they do not interfere with the operation of the machine tool. This is true with both automated and conventional machine tools. The overall height of the clamping devices should always be as low as possible.

When stud and nut arrangements are used, they must not extend too far above the top of the clamp. Fig. 2-15 shows a few general considerations for strap clamp arrangements. As a rule, a stud should not extend more than 2 to 3 threads above the top of the nut, as shown in part A. If the height of the nut and stud in a strap clamp assembly causes a problem, as shown in part B, consider a bolt instead of a stud and nut. But if a stud and nut arrangement must be used, as in part C, replace the standard flat clamp strap with a gooseneck style to prevent interference with the movements of the machine.

As a final check, before using the workholder, always test the setup manually. This is done with the tool set well above the workpiece to ensure everything is clear. Look for areas where the workholder might interfere with the machine tool. Always check the complete machining cycle. With milling machines, look for problems with the column, quill, arbors, or the table travel. With a lathe, look for interference with the cross slide or the compound rest.

### General Construction Principles

Tool bodies are the main element in any jig or fixture. The tool body provides the mounting area for all the locators, supports, clamps, and other components that position and hold the workpiece. The design and construction of a tool body are usually decided by the workpiece, the operations to be performed, and the expected life of the workholder. Economy is another factor that should also be considered in the design decision.

Tool bodies are usually made in three basic types or forms—cast tool bodies, welded tool bodies, and built-up tool bodies (Fig. 2-16). Each tool body can be adapted for any workpiece; but in some applications, one is preferred over another. The first step in deciding which tool body to use is to understand the strengths and weaknesses of each variation.

*Cast Tool Bodies*

Cast tool bodies are made in a variety of styles. The most common material used for cast tool bodies is

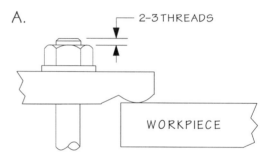

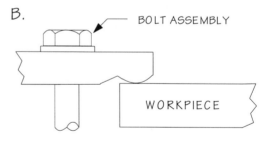

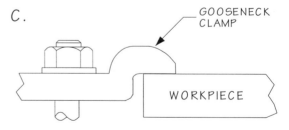

**Fig. 2-15.** Strap clamp arrangements should be designed so they do not interfere with the machine tool.

# Basic Workholding Principles

cast iron or cast aluminum. Other cast materials sometimes used for cast tool bodies are low-melt alloys and epoxy resins. Low-melt alloys are bismuth alloys with melting temperatures from 117°F to 440°F (depending on the composition). These materials are better for specialized workholder elements rather than for complete tool bodies. Low-melt alloys, unlike other cast materials, are also reusable. Epoxy resins are another material found in special locating and clamping elements. But, depending on the size of the workpiece, these materials may also be used to cast complete tool bodies.

Cast tool bodies offer excellent dimensional stability and distribution of material. They may also be cast into complex and detailed shapes that usually require less secondary machining. Cast materials also provide good vibration dampening qualities. They are most often used for workholders that are relatively permanent and not subject to drastic changes. The major drawbacks to cast tool bodies are their inability to be easily changed or modified, and the higher initial cost of the castings.

### Welded Tool Bodies

Welded tool bodies are also made from a wide variety of materials. The most common are made of steel and aluminum. Welded tool bodies are most often used for workholders of a semi-permanent nature that do not require extreme precision. Although there are exceptions, welded tool bodies are more common for workholders intended for roughing rather than for finishing operations.

Welded tool bodies are inexpensive to build and are generally easy to modify. Unlike cast tool bodies, welded tool bodies can be put together quite rapidly. They are also durable and rigid and provide an excellent strength-to-weight ratio. Their major problem is heat distortion. Depending on the materials and methods, the heat of welding can cause distortion of the welded pieces. Occasionally, secondary machining operations are needed to remove this distortion. Another problem area with welded tool bodies is with dissimilar materials. When a steel block, for example, is added to an aluminum tool body, it is attached with threaded fasteners rather than welded to the body.

### Built-Up Tool Bodies

Built-up tool bodies are the most common form of tool body. These tool bodies are very easy to build and normally require the least amount of time between design and finished workholder. Built-up tool bodies are usually made of individual elements, assembled with screws and dowel pins. Machined parts are the most common elements for this type of tool body.

The built-up tool body is very easy to modify to suit changes in the part design. Like the welded type body, these tools are durable and rigid. They have a very good strength-to-weight ratio. Depending on the complexity of the design, the built-up type can also be the least expensive type to construct. Built-up tool bodies are well suited for a wide variety of different applications. Typically, built-up tool bodies are intended for

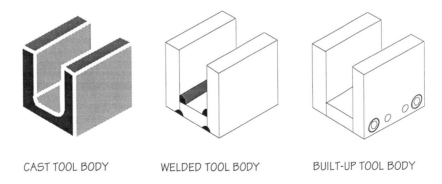

**Fig. 2-16.** Tool body variations.

workholders for precision machining operations, inspection tools, and some assembly tools.

## Designing Tool Bodies

One way to reduce the amount of time and labor involved in designing and building a workholder is with preformed materials or commercial tool bodies. Both preformed materials and commercial tool bodies are available in a variety of materials, sizes, and forms for a wide range of applications. The following is a description of the more common variations of preformed material for building jigs and fixtures.

### Precision Ground Flat Stock

Precision ground stock is readily available in a variety of size/material combinations. The most common forms are square and rectangular sections. They are usually available in three material types: oil-hardening, air-hardening, and low-carbon. The oil-hardening types are usually O1 tool steel, the air-hardening are A2 tool steel, and the low-carbon are 1018/1020 carbon steel.

Although the available sizes may vary, the most common types come in sizes ranging from 0.016" × 0.016" to 2.000" × 6.000". Generally, these sections are available in 36" lengths in either a ground-to-an-exact-size or ground-oversize condition. The size tolerances for most of these materials are ±0.001"/foot. These tolerances apply to the size, flatness, squareness, and parallelism of the section.

To use these materials, simply cut off the amount of material needed and store the rest of the bar. These materials are well suited for base elements, locators, riser blocks, and almost any fixturing element.

### Drill Rod

Drill rod, like precision ground stock, is available in a wide range of sizes. But the most common variation for workholding applications is the drill blank. Drill blanks are available in standard sizes from 0.013" to 0.500" in the number, letter, and fractional sizes corresponding to the standard drill sizes. These elements are well suited for locators, supports, arbors, or similar applications where a cylindrical form is required.

### Cast Bracket Materials

Cast bracket materials are available in a variety of shapes and sizes (Fig. 2-17). The two most common forms of cast bracket materials are cast iron and cast

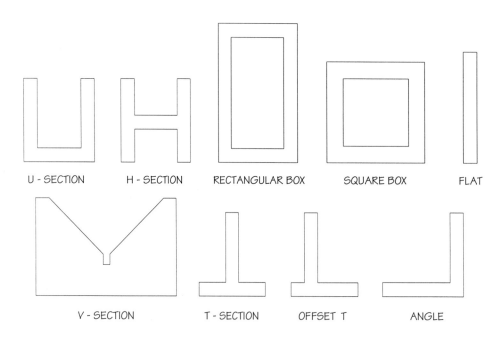

Fig. 2-17. Cast bracket materials.

# Basic Workholding Principles

aluminum. They are usually found in standard lengths of 25.00" with tolerances of ±0.005"/foot on all working surfaces. Rather than being used as accessory items, most cast elements are designed as major structural elements of the workholder. Depending on the complexity of the workholder design, it may be possible to build a complete workholder by combining different sections.

## Structural Sections

Structural sections, such as those shown in Fig. 2-18, are also well suited for a variety of workholder applications. These sections are readily available in a variety of sizes, shapes, and materials. The most common materials used for workholders are steel, aluminum, and magnesium. Although they lack the precision and accuracy of other preformed materials, they can be employed as either individual elements or complete portions of workholders where high precision is not required. When higher precision is required, each of these structural sections may be machined to suit the specific requirements.

## Commercial Tool Bodies

Commercial tool bodies are made in a variety of forms and types to accommodate a variety of tooling applications. Two of the more common styles of commercial tool bodies are the channel body and fixture base (Fig. 2-19). Both styles are available in several different sizes to suit a range of different workpiece sizes.

The principal advantage of preformed materials is the reduced machining required to complete these elements. Since these elements are made to the approximate shape (or form), only minimal machining is required to finish the parts. This greatly reduces the time and labor needed in building the complete jig or fixture.

## General Design Considerations

When selecting the materials and designing a tool body, there are a few general considerations to keep in mind. Since the tool body is usually the largest single element in any workholder, it also offers the greatest potential for saving money. When possible, select the materials for the tool body that offer the best economy. Use low-cost materials, such as 1020 carbon steel, for areas where only mass is required. Do not use tool steels unless the fixturing element will be hardened.

When tool steel is specified, using an A2 tool steel usually offers the best economy. Other tool steels, such as O1, although initially less expensive, do cost more to process than A2, or other air-hardening tool steels. Try

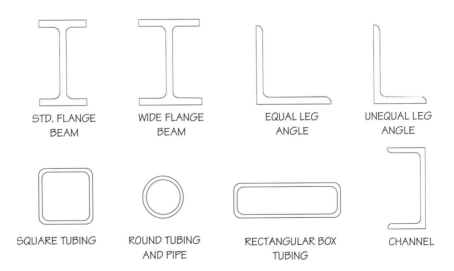

**Fig. 2-18.** Structural sections.

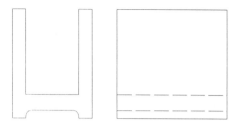

PRECAST CHANNEL BODY

PRECAST FIXTURE BASE (SQUARE)

PRECAST FIXTURE BASE (RECTANGULAR)

**Fig. 2-19.** Commercial tool bodies.

to avoid specifying D2 or other wear-resistant tool steels since these materials are considerably more difficult to work with, and can add substantially to the processing costs.

The tool body must be designed to be sturdy enough to maintain the alignment of the various locators, clamps, and other elements. Likewise, the tool body must not bend or distort during machining. To prevent distortion, it is usually more economical to specify a thicker piece of soft 1020 carbon steel than a thinner piece of hardened tool steel.

# CHAPTER THREE

# Workholding Options and Economics

# Workholding Options and Economics

Once the proper ground rules are set, designing economical, efficient, and cost-effective workholders is actually rather simple. The ground rules, like many other design factors, are a mix of practical considerations, sound design practices, and common sense. So, instead of an academic exercise or a form of technical magic, design economy, like setup reduction, is a process of matching workpiece requirements with the financial realities of the manufacturing budget.

Balancing design considerations with budget restrictions may at first seem difficult. However, in practice, economic design not only tends to make a workholder less costly, but often dramatically improves its efficiency and operation as well.

The two guiding principles to reducing setup expenses and designing cost-effective workholders are *simplicity* and *economy*. By expanding on these two themes, the basic principles of economic design can be explored and analyzed. Applying these basic principles not only reduces the cost of workholders, but many times improves their overall operation.

## Workholding Options

No single type or style of workholder is completely suitable for every workpiece. Just as there are an infinite number of different workpiece and product configurations, there are also a variety of workholder options. When selecting a particular style of workholding device, the designer must concentrate on a number of variable factors. Whereas a particular workholder design is determined by the workpiece characteristics, the style of workholder is usually influenced by the production factors.

Before beginning a workholder design for a specific workpiece, the designer must first decide which style of workholding device to use. The designer typically has three choices: temporary workholders, dedicated workholders, and modular workholders. In some situations, any of these three forms of workholders may be appropriate and will work equally well for a particular job. More often, however, only one will be the best choice. No single style of workholder is fundamentally better than another—the best choice depends on the specific application.

Several factors are evaluated when selecting the style of a workholder. However, as shown in Fig. 3-1, the size of the production run and the number of times the production run is expected to repeat are usually the major factors. Other considerations include the expected length of production, amount of available time and money, and when the tooling is needed. Other factors may also be added to this list, but the basic point is simply that the size of the production run, although important, is not the sole factor in selecting a workholder.

Most shops have applications where all three styles of workholders are ideal. Rather than completely different forms of workholders, each of the fixtures is simply an alteration intended to suit specific production requirements. To apply any of these workholders most efficiently, you should have a working understanding of the characteristics and features of each style.

### Temporary Workholders

Temporary workholders are the simplest and least expensive style of fixturing devices. This category of workholder includes the wide array of standard commercially available clamps, vises, chucks, collets, and similar components. These elements are generally reusable and serve as general-purpose fixturing components. Strap clamps are a form of temporary workholder frequently used for table-top workpiece setups.

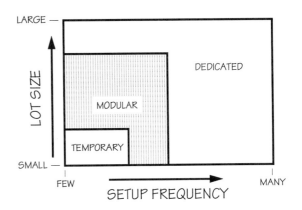

**Fig. 3-1.** The style of workholder for any job is typically decided by the size of the production run and the frequency of production.

The most common style of temporary workholder is the vise. Standard machine vises may be used with standard jaws. When required, special vise jaws may be made to fit the shape of the workpiece. Lathe chucks and collets are other examples of temporary workholders that may also be used in their standard configuration or with special gripping elements made to suit the workpiece.

Temporary workholders are often employed for prototype or other one-of-a-kind jobs. Depending on the particular workpiece, they may also be satisfactory for limited or short production runs where a dedicated workholder is not economical. Although temporary workholders represent a smaller initial investment, they are often inadequate or unacceptable for complex parts or high-volume production.

*Dedicated Workholders*

Dedicated workholders are special-purpose fixtures specifically designed and constructed for a single workpiece or family of similar workpieces. These workholders are usually the most efficient and the most expensive workholder variation. As such, dedicated workholders are typically the most common form of workholder used for manufacturing. Dedicated workholders are constructed from a variety of standard and custom-made elements to meet the specific requirements of the workpiece. Although dedicated workholders are the most expensive, they are the best choice for high-volume or repetitive production runs.

Dedicated workholders are designed as permanent tools intended for mass production of many parts. They are most often intended for a single operation on one particular part. This specialization, while expensive for only a few parts, is usually the most economical for larger part volumes. High-volume production allows more time and money to be spent building these workholders. Limited production runs require the cost of the workholder to be absorbed by a smaller number of workpieces. With dedicated fixturing, however, the additional cost of these workholders is justified through the savings achieved over many more workpieces.

*Modular Workholders*

Modular workholders are a unique fixturing variation which combines the better features of both temporary and dedicated workholders. A modular fixture is a special-purpose workholder assembled from a variety of general-purpose components. Modular workholding is not a completely new form of fixturing; instead, it is actually a simplification of dedicated fixturing methods. Modular fixtures are constructed with virtually all the accuracy, detail, and capability of dedicated workholders, but at a cost comparable to temporary workholders.

Because of the way it is constructed, the overall cost of a modular workholder is considerably less than a dedicated alternative. Dedicated workholders are constructed with custom-made elements specifically made to fit the requirements of a workpiece. Modular workholders, however, are assembled from a series of general-purpose elements that are both reusable and universal. Once a modular workholder is no longer needed for production, the fixture may be disassembled, and the individual components may be used again for other workholders. Modular component workholding systems, or flexible fixturing systems as they are also called, can be used for almost any kind of part. In most cases, the only limitation with this form of workholder is the imagination and creativity of the designer.

Modular component workholding systems are available in three basic styles, or types: subplate systems, T-slot systems, and dowel pin systems. In addition, within each of the general categories, there are also several variations. Some modular system manufacturers use different mounting patterns as well as slightly different accessories. The specific type of modular workholding system best suited for any company normally depends on the individual requirements and the tasks required for the tooling. The following is a brief description of these three styles of modular fixturing systems.

**Subplate Modular Fixturing Systems:** The most basic and elementary type of modular fixturing system is the subplate system. These systems are composed of a series of structural fixturing elements that incorporate a variety of commercial locators and clamps to fixture the workpieces. The most common variations of these subplate structural elements are: plain flat plates, angle plates, multi-sided prisms, riser blocks, parallels, and similar components (Fig. 3-2).

# Workholding Options and Economics

Subplate systems, depending on their design, may have either a dowel pin arrangement or T-slots to provide mounting points for the additional attachments and accessories. In addition to being used for some relatively simple workholders, subplate workholders are also used for mounting other workholders to machine tool tables.

**T-Slot Modular Fixturing Systems:** The T-slot type system is the oldest form of modular fixturing. These workholders use a series of precisely machined baseplates, mounting bases, riser blocks, and similar elements to construct workholders. Each of these structural elements is made with a series of machined and ground T-slots (Fig. 3-3). The T-slots act as mounts to attach the additional fixturing components and accessories. Unlike subplate systems, which contain only the major structural elements, T-slot systems also offer a wide array of locating, supporting, and clamping devices as part of a total fixturing system.

T-slot systems are usually quite accurate and are normally made to very close tolerances. The principal advantage of the T-slot type system is its overall strength and adaptability. The basic T-slot arrangement is considerably stronger and more durable than the dowel pin design. But, if a T-slot is damaged, it can also be more difficult to repair. The basic design of a T-slot, like a machine tool table, is very adaptable and permits the components to be positioned very easily. For the most part, workpieces are easier to clamp when the workholding elements are located in T-slots.

The principal drawback to this form of modular fixturing system is in its limited number of registration points. As shown in Fig. 3-4, a typical T-slot system contains a series of T-slots that are parallel and perpendicular to each other. The only fixed registration points are where these T-slots cross each other. These registration points are the only fixed reference points for locating other fixturing elements on the baseplates

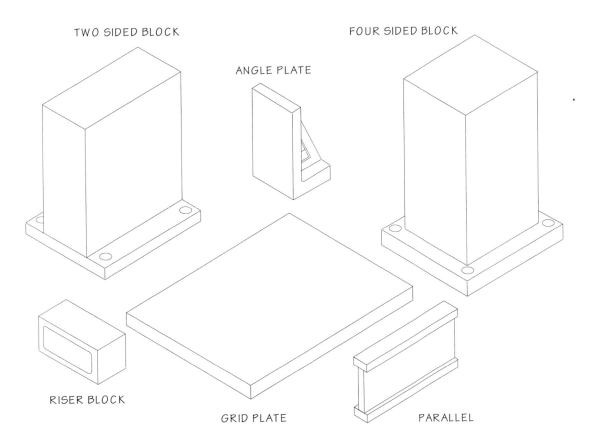

**Fig. 3-2.** Subplate structural elements.

or other components. When a workholder is assembled the second or subsequent time, these registration points assure repeatability from one tool to the next. When components are located away from these registration points, more care must be exercised to ensure that the components and elements are properly positioned on the base elements.

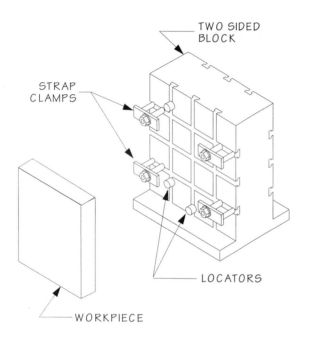

Fig. 3-3. T-slot modular fixturing system.

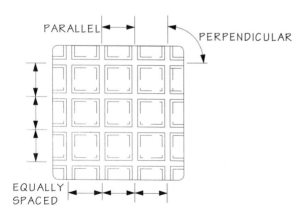

Fig. 3-4. T-slot systems have parallel and perpendicular T-slots. The points where the T-slots cross each other establish the fixed registration points.

***Dowel Pin Modular Fixturing Systems:*** The dowel pin style modular fixturing system is the newest form of modular workholder. These systems are very similar in their basic design to the T-slot type workholders. The overall size, capabilities, and range of available components are very similar between these two systems. The major difference between the two styles is in the methods used to attach components. Instead of T-slots, the dowel pin design has a grid pattern of holes to locate and mount the other accessories (Fig. 3-5).

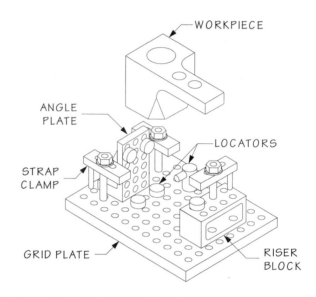

Fig. 3-5. Dowel pin modular fixturing system.

The major advantage of a dowel pin type system is the large number of registration points (Fig. 3-6). Each dowel pin hole acts as a reference point, and permits virtually automatic positioning of the components from one tool to the next. If a workholder must be built more than once, the dowel pin arrangement makes locating and positioning the components much faster and easier. Likewise, since all the major fixturing components are located with dowel pins, the positional accuracy of the location is also ensured.

Dowel pin systems are available in two general styles (Fig. 3-7). Depending on the specific system, the holes in a dowel pin system may have either an alternating pattern or a combination design. The alternating style, shown in part A, has alternating tapped holes sepa-

# Workholding Options and Economics

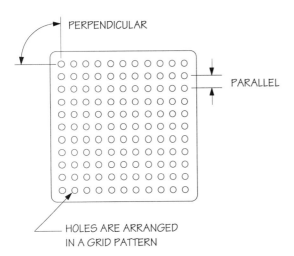

**Fig. 3-6.** The dowel pin design has a grid pattern of holes. Each dowel pin hole acts as a fixed registration point.

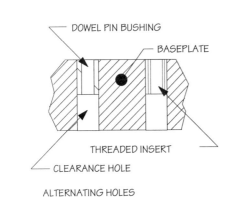

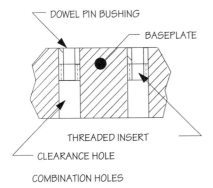

**Fig. 3-7.** Dowel pin systems are available with two general styles of mounting/locating holes.

rated with dowel pin holes. In this arrangement, both dowel pins and screws locate and mount the components. The second style, shown in part B, combines both the locating and mounting functions in each hole by mounting a locating bushing on top of a tapped hole.

The alternating hole arrangement typically uses standard dowel pins and socket head cap screws to locate and attach the various components. The combination hole style typically requires special-purpose locating screws. These screws both locate and attach the components. Virtually all dowel pin modular fixturing systems use replaceable dowel pin bushings and thread inserts. This design allows easy replacement of any damaged elements.

The main disadvantage of the dowel pin type system is in the methods used to clamp a workpiece. Unlike the infinite adjustability of the T-slot, dowel pins have a fixed positional relationship to each other. Due to the spacing of the threaded holes, the components and arrangements employed for clamping some workpieces are often more awkward and cumbersome than those used with the T-slot systems.

Regardless of the drawbacks of modular workholding, any modular workholding system is generally superior to dedicated workholders for a majority of applications. Just the amount of time that can be saved will often justify the cost of a modular fixturing system. The average modular component workholder can be designed and built in approximately 5% to 20% of the time required for a comparable dedicated workholder.

## Workholder Selection

As mentioned earlier, there is no truly universal workholder available in manufacturing today. Likewise, no single style of fixturing is the complete answer to every workholding requirement. The best way to approach workholding is with a combination of temporary workholders, dedicated workholders, and modular workholders. Using all three forms of fixtures allows each style to be applied where it is best suited and appropriate. Furthermore, applying each form of workholder where it is most beneficial offers the most efficient and economic approach to workholding.

When selecting a specific form of workholder for any particular workpiece, several factors, as well as a few general workholder design guidelines, should be considered. Although not all of these general considerations are appropriate for each workpiece or workholding situation, the effect of each should be examined to develop the best workholding alternatives.

To reiterate, as a rule, most workholder selections are based on many general factors. The more common considerations are the type of workpiece, the type of operations to be performed, the number of workpieces, the number of production runs, and the expected life of the product or workpiece.

For applications where large numbers of workpieces are run on a regular basis, dedicated workholders are usually the most logical choice. Where a moderate to small number of workpieces are produced on an irregular basis, modular workholders may be a better alternative. Likewise, for small runs or one-of-a-kind simple parts, temporary workholders such as vises, collets, or chucks may offer the most economy and efficiency.

Dedicated workholders should be used for those workpieces that are in current or long-term production. Modular fixturing is best suited for parts produced on an as-needed basis. Although dedicated workholders may handle the bulk of everyday production, modular workholders can be employed on an as-ordered basis to make small numbers of special parts. In addition to limited or short-run production, modular workholders are also good for prototype or experimental fixturing, or as temporary replacements for dedicated workholders that must be pulled out of production for routine maintenance or to repair damage.

Modular fixturing elements and components may also be used in building dedicated workholders. Depending on the requirements of the fixture and the workpiece, both custom-made and modular components can be combined to construct a workholder. Although dedicated workholders are most often made with specialized, custom-made parts, where practical, modular elements may be used in place of some specialized elements.

Likewise, when assembling modular workholders for some complex or detailed workpieces, some special custom components may be required. Here the custom-made components are made to suit both the workpiece and the modular workholder. This approach combines the best features of both fixturing styles. One other approach to keep in mind when working with modular fixturing is simply leaving the modular workholder assembled between production runs. While modular workholders are commonly disassembled after each use, they can be left together. In this case, the workholder functions as a dedicated workholder built with modular elements. When building workholders, consider all three styles of fixtures and select the most effective, efficient, and economical.

## Modular Fixturing Versus Dedicated Fixturing

When selecting a workholder for a particular job, the workpiece and the production requirements usually dictate the specific style of workholder. Sometimes this choice is rather easy to make. For example, the choice between a temporary and dedicated workholder is usually not very difficult to decide. However, when choosing between a dedicated and modular workholder, the selection may not be quite as obvious. The general guidelines for workholder selection mentioned earlier help simplify this decision. However, to further evaluate and define this selection, there are a few additional economic factors that should be considered.

### Design Economics

Neither dedicated nor modular workholders are better suited for every possible fixturing task. When choosing, one place to begin the analysis is with the economic considerations of each style of workholder. The following design areas are those that should be examined to ensure a valid comparison between a dedicated and modular workholder.

**Design Time:** The time required to design either dedicated or modular workholders is basically the same with manual drawing methods. With CAD, however, a modular fixture can take considerably less time. Almost all the manufacturers offer modular component databases. These databases allow the designer to select a component and drop it into the drawing, rather than drawing the complete detail, as required for dedicated fixtures.

***Material Cost:*** The material cost for a dedicated fixture is simply the total cost of all the materials, hardware, and components used to construct the workholder. With modular workholders, however, this value is the total cost of the components used to assemble the fixture. This is not really a direct comparison, since the units compared are only similar—not identical. For example, if the same element were constructed as a dedicated component, made in-house, it might be less expensive than a similar modular element. But when you consider that the modular element can be reused, then the cost, on a per-use basis, of the modular element is usually considerably less than that of the dedicated elements.

***Fabrication Cost:*** The cost of fabricating the fixturing elements is a major factor in the total cost of constructing a workholder. In fact, this is usually the most expensive element in building a dedicated workholder. The fabrication involves all the machining required to make each element. It also includes the assembly of the completed workholder. Modular component workholders, on the other hand, are assembled from a series of premade fixturing elements. In most cases, no machining operations are required for a modular workholder. The fabrication costs involved with modular fixtures are only a fraction of the cost normally associated with dedicated workholders.

***Inspection and Testing Expenses:*** The processes used to test and inspect both dedicated and modular workholders are identical. So, the cost for inspecting and testing either style of workholder is essentially the same. Both types of workholders should be inspected and tested before they are released to production.

***Tool Storage and Maintenance:*** The cost of storing dedicated jigs and fixtures is a frequently overlooked expense in many companies. The costs include the cost of the space required to inventory the tool, the record keeping necessary to keep track of the workholders, and the expense of retrieving and replacing the tool each time it is used.

Maintenance is another costly element in storing dedicated workholders. The average dedicated workholder requires approximately two hours of maintenance each time it is used. The expenses include the time required to retrieve the workholder, and the time involved in validating the accuracy of the tool before it is put into production. In addition, when the cost of periodic inventories and preventative maintenance for these stored tools is considered, the cost can be significantly more.

Modular workholders, on the other hand, have very few of these liabilities and additional expenses. They do not require a warehouse or other large storage area. Instead, modular workholders are often stored right in the toolroom. When a modular workholder is built, the only documentation required consists of a series of photographs, or videotapes, and a parts list. The photographs, videotapes, and parts list are the only records of the workholder that need to be kept for future use.

Once a modular workholder is no longer needed for a particular job, the workholder is disassembled. The individual parts are returned to their storage area in the toolroom, ready for the next workholder. There is little or no maintenance involved with modular workholders since each fixture is built and verified as needed. There is no need to verify the tool as an independent step.

*Estimating the Cost of Workholders*

The first step in estimating the cost of any workholder is to make sure the correct method of estimating is used for the workholder. As a general rule, dedicated workholders are expensed. Modular workholders are amortized. The major reason for this difference is that dedicated workholders, once made, cannot be used for anything but the part for which it was initially made. Modular workholders, on the other hand, are usually assembled for a workpiece and disassembled at the end of the production run. The individual modular elements and components can then be reused to assemble other workholders for completely different workpieces.

When workholder expenses are calculated, the cost of a dedicated workholder should include the cost of the material, hardware, and components as well as the labor expense in machining the individual elements of the workholder. Modular workholders, however, are assembled from a series of elements that do not require these calculations. The cost of the modular elements is amortized over a preset life that is determined by the user.

The simplest way to find the cost of a dedicated workholder is to first identify all material, hardware, and commercial fixturing components required for the proposed workholder. The individual cost of each of these is then added to find the total cost of materials. An estimate of the time required for designing, machining, and assembling the elements is then prepared. This estimated time value is then multiplied by the shop labor rate to find the total labor expense. This labor expense is then added to the material cost to determine the estimated cost of the dedicated workholder.

To estimate the cost of a modular fixture, first find the cost of the individual components. There are two methods used to find this value: the approximate cost method and the exact cost method. The process used normally depends on the desired accuracy of the estimate. Once the cost of the individual elements is found, the total cost of the workholder is found by adding the cost of all the components actually used to construct the workholder.

The simplest, but least accurate, calculation is the approximate cost method. With this method, the total cost of the modular fixturing system is divided by the number of components contained in the system. While not as precise as the exact cost method, since the baseplates cost more individually than the studs or nuts, it will give a reasonably close estimated cost value. So, if the modular fixturing system cost $45,000 to buy and contained 900 individual parts, the cost per part would be $45,000/900 = $50. To find the total cost of the fixture, this $50 value would simply be multiplied by the number of components used for the workholder.

If a more accurate estimate is necessary, the exact cost method may be used. This method, although more accurate, is also more time consuming. First, the specific components used for a workholder are identified. Then, using a price sheet, the individual cost of each component used for the workholder is added to find the total cost of all the components.

Once the cost of the components used to construct the workholder is found, the amortized value is then calculated. This amortized value is simply the cost of the workholder applied on a per-use basis. The factors used here are determined by the individual user, and reflect the time frame required to pay for the complete modular workholding system. For example, if the company wanted to pay for the system with 20 workholders, the amortized value is found by dividing the cost of the components by 20. A more realistic value, however, is 100 uses. This value equals 10 fixtures per year for a life of 10 years. This is actually quite conservative since any modular system, with just minimal care, has a life expectancy of 20 to 30 years. The values used for these calculations, expressed as formulas, are shown in Fig. 3-8.

Amortized Cost Per Element:

$$a = \frac{IC}{(n \times U \times L)}$$

Amortized Cost Per Tool:

$$A = a \times N$$

Where:
- $a$ = Amortized cost per element
- $IC$ = Initial cost of the system
- $n$ = Number of parts in the system
- $U$ = Uses per year
- $L$ = Life expectancy of the system
- $A$ = Amortized cost per tool
- $N$ = Number of parts used for a tool

**Fig. 3-8.** Calculating the amortized value of a modular workholder.

## Workholding Options and Economics

*Analyzing Tooling Expenses*

The simplest way to compare dedicated and modular workholders is by analyzing and comparing the cost of the various economic factors. To ensure an accurate comparison, the analysis must be made by evaluating both styles of workholders designed for the same workpiece. Fig. 3-9 shows a sample comparison of a dedicated and modular workholder for the same workpiece. Once the relative cost of each type of workholder is determined, the next step is to calculate the break-even point for both types of workholders.

The break-even point is a calculation that helps determine whether a modular or dedicated workholder should be used for a specific task. The cost of a dedicated workholder, once built and paid for, should not be calculated each time the workholder is used. However, a modular workholder does require reassembly each time it is used. Therefore, this cost must be included in the calculation. By making a break-even calculation, the relative cost of both workholder styles can be determined on a per-use basis. So, while a modular workholder may have an initially lower cost than a dedicated workholder, each time the modular workholder is configured, a cost is incurred. The break-even calculation allows the designer to determine the number of times a dedicated workholder needs to be used before it is more economical than a modular workholder. As shown in Fig. 3-10, this calculation involves calculating the cost of both workholders, for each time they are used. The end result is the number of times a dedicated workholder needs to be reused before it is more economical than a modular workholder for the same workpiece.

### Cost Comparison

*Conventional Dedicated Fixture*

| | |
|---|---|
| Planning & Design (11 hrs @ $40.00) | $440.00 |
| Material Cost | 186.00 |
| Fabrication & Assembly (47 hrs @ $40.00) | 1880.00 |
| Testing & Inspection (3 hrs @ $40.00) | 120.00 |
| Tool Storage Expense (per year) | 80.00 |
| Total Expenditure | $2706.00 |

*Modular Component Fixture*

| | |
|---|---|
| Planning & Design (4 hrs @ $40.00) | $160.00 |
| Material Cost | 0 |
| Assembly (5 hrs @ $40.00) | 200.00 |
| Disassembly (1 hr @ $40.00) | 40.00 |
| Amortization of Elements (48 pcs @ $43.00/100) | 20.64 |
| Testing & Inspection (3 hrs @ $40.00) | 120.00 |
| Tool Storage Expense (per year) | 0 |
| Total Expenditure | $540.64 |

Total savings of $2165.36

(Roughly 20% of dedicated tooling cost)

**Fig. 3-9.** Comparing the costs of a dedicated and modular workholder built for the same workpiece.

### Break-even Calculations

*Cost Per Use*

| | Dedicated | Modular |
|---|---|---|
| Retrieval & Testing | $80.00 | $ 0 |
| Assembly | 0 | 200.00 |
| Tool Testing & Inspection | 0 | 120.00 |
| Disassembly & Storage | 80.00 | 40.00 |
| Cost Per Use | $160.00 | $360.00 |

*Break-even Calculation*

$$\frac{\text{Initial Cost of Dedicated} - \text{Initial Cost of Modular}}{\text{Cost per use of Modular} - \text{Cost per use of Dedicated}}$$

$$\frac{\$2706.00 - 540.64}{\$360.00 - 60.00} = \frac{\$2165.36}{\$200.00} = 10.83 \text{ Fixtures*}$$

\* If the setup will be run more than 11 times, dedicated fixturing would be the more economical. If the setup is run less than 11 times, then modular fixturing would be the most economical.

**Fig. 3-10.** Calculating the break-even point.

# PART TWO
## Setup Reduction Techniques

# CHAPTER FOUR

Techniques
for
Locating

# Techniques for Locating

*Description of Problem/Requirement:*

Dowel pin locators do not offer desired locating flexibility.

*Suggested Solution:*

Replace dowel pins with commercial locating pins.

*Source:*

Various jig and fixture component manufacturers.

*Old Method:*

Dowel pins are often used as locators for some workholders. While sometimes adequate, standard dowel pins do have serious limitations when used for locating in jigs and fixtures.

- Dowel pins have a limited range of standard diameter sizes.

- Dowel pins have only one shape. They lack the variations required for some workpieces and workholding setups.

- The cylindrical form of the dowel pin does not offer an easy method of relieving the locator. Any form of groove cut into the dowel for this relief can weaken the dowel.

- Dowel pins often lack the appropriate hardness for locating. Most are either too hard or too soft for locational applications.

- The end shape of most dowel pins makes loading hole-mounted workpieces very difficult.

*New Method:*

Commercially available locators are better suited for locating than dowel pins.

Standard locators are available in either plain or shoulder styles with a variety of end shapes.

Plain Type

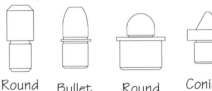

Round Pin    Bullet Nose    Round Nose    Conical Pin

Shoulder Type

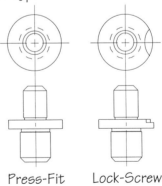

Press-Fit Type    Lock-Screw Type

| |
|---|
| *Description of Problem/Requirement:* |
| Tool bodies must be replaced when dowel pin locators wear out. |
| *Suggested Solution:* |
| Replace dowel pin locators with a replaceable locator arrangement. |
| *Source:* |
| Various jig and fixture component manufacturers. |

| Old Method: | New Method: |
|---|---|
| A plain dowel pin was pressed into the tool body to act as a locator. When the dowel pin was damaged or became worn, it was replaced with another dowel pin. When the mounting hole became too large to accurately hold the dowel pin, the complete tool body was replaced. | Replaceable locator arrangement is used instead of the press-fit dowels. If the locator pin is damaged or wears out, it is simply removed and replaced with a new locator pin. |

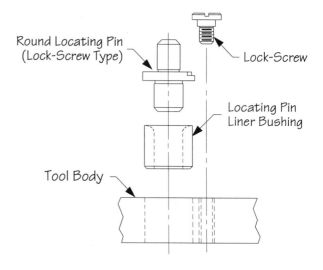

# Techniques for Locating

**Description of Problem/Requirement:**

Provide an accurate and positive method to locate and align workholder elements.

**Suggested Solution:**

Incorporate a locator/bushing arrangement into the workholder design.

**Source:**

Various jig and fixture component manufacturers.

**Old Method:**

A plain dowel pin is pressed into the bottom element of the workholder and aligned with a drilled and reamed hole in the top element. Although adequate for very short production runs, after repeated use the hole wears out and the alignment accuracy is lost. When this occurs, the hole is bored oversize and a custom bushing is fabricated and installed.

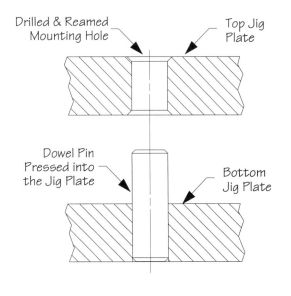

**New Method:**

A commercial locator/bushing arrangement is used instead of the dowel pin method. Since both the locating pin and bushing are hardened, their service life is greatly extended. Also, if either (or both) the locator pin or bushing is damaged or wears out, it is simply removed and replaced.

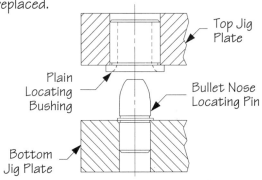

### Locator/Bushing Variations

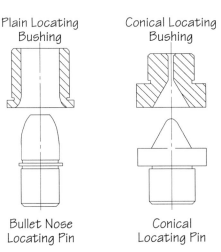

## Techniques for Locating

**Description of Problem/Requirement:**

Simplify workpiece mounting by making the process faster and easier.

**Suggested Solution:**

Reduce the contact area between the workpiece and the locators.

**Source:**

Various jig and fixture component manufacturers and in-house fabrication.

**Old Method:**

The contact area between the workpiece and locator is an important consideration in locating. The greater the contact area, the more difficult the workpiece is to load. Ideally, the contact area should be between 1/8 and 1/2 the workpiece thickness.

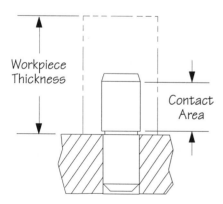

**New Method:**

### Standard Commercial Diamond Pin

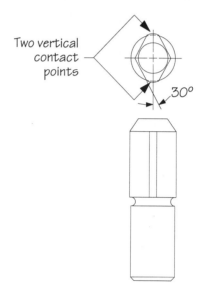

### Relieved Locator Variations

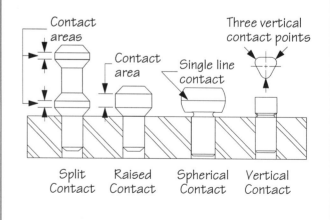

# Techniques for Locating

*Description of Problem/Requirement:*

Reduce the tendency for the workpiece to jam on hole-mounted locators.

*Suggested Solution:*

Design a locating arrangement that uses a diamond pin locator.

*Source:*

Various jig and fixture component manufacturers.

*Old Method:*

Using two dowel pins for a hole-mounted locating arrangement can create problems in loading and unloading the workpiece. The added contact area between the dowel pins and the workpiece causes the workpiece to bind or jam on the locators.

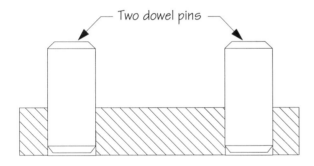

Two dowel pins

*New Method:*

Using one round locator and one diamond pin locator for hole-mounted locating arrangements makes the loading and unloading operations much easier and reduces the tendency for the workpiece to bind or jam on the locators.

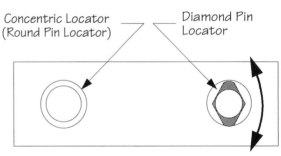

Diamond pin locator prevents rotation around the concentric locator.

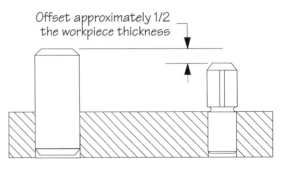

Diamond pin is installed slightly lower than the concentric locator to make loading easier.

**Description of Problem/Requirement:**

Reduce the tendency for the workpiece to jam on hole-mounted locators.

**Suggested Solution:**

Design a locating arrangement using two diamond pin locators.

**Source:**

Various jig and fixture component manufacturers.

**Old Method:**

The preferred method of applying a diamond pin locator is as a radial locator to prevent rotational movement of a workpiece about a round locator.

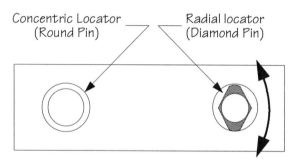

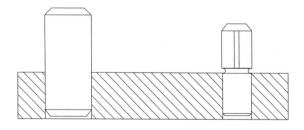

**New Method:**

Where the locational tolerances are not as tight, however, two diamond pins may be used to locate a workpiece. Here the pins are positioned at 90° from each other and will restrict the movement of a workpiece in both directions. This makes loading and unloading the workpiece much easier. But this method should only be used where there is sufficient locational tolerance.

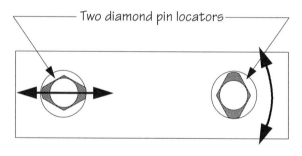

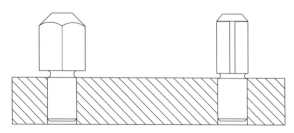

Each diamond pin locator prevents workpiece movement about the other diamond pin.

# Techniques for Locating

*Description of Problem/Requirement:*

Furnish maximum locational accuracy while reducing locator contact.

*Suggested Solution:*

Apply specialized forms of relieved locators to position the contact points where needed.

*Source:*

In-house fabrication.

*Old Method:*

To minimize locating problems, the contact area between the locator and workpiece should be minimized. However, the location of the contact area is just as important to proper location as the amount of the contact.

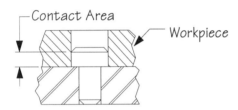

The contact area, although reduced, is positioned at the bottom of the hole, against the tool body.

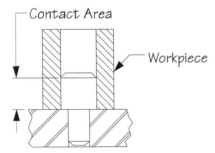

With thicker workpieces, placing the contact area completely at the bottom of the hole can sometimes cause locational problems since the top area of the workpiece is not supported.

*New Method:*

Specialized forms of relieved locators may be used to position the contact area exactly where needed to suit the requirements of the workpiece.

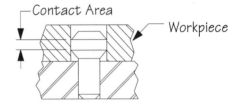

Raised contact locator elevates the contact area of the locator off the tool body and engages the workpiece near the center of its thickness.

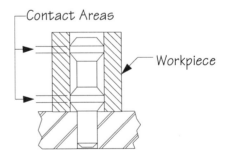

Split contact locator divides the contact area over the thickness of the workpiece. Although the workpiece contact is reduced, the support is maintained by locating the workpiece at both the top and bottom of its thickness.

# Techniques for Locating

**Description of Problem/Requirement:**

Provide a hole-mounted locator that cannot jam or bind in the workpiece.

**Suggested Solution:**

Utilize spherical locators to prevent jamming.

**Source:**

In-house fabrication.

**Old Method:**

Standard locating pins, such as dowel pins or commercial locating pins, have a cylindrical form. If a workpiece is loaded on these pins, the mounting hole will bind unless the center lines of the hole and locator are precisely aligned. This is caused by the effect of the hypotenuse of the triangle formed by the relationship of the workpiece and locator when they are not properly aligned. The diameter of the pin ($d$) is the true diameter of the pin. However, if the workpiece is tilted, the relationship of the pin to the mounting hole ($D$) has an elliptical rather than a round form. This elliptical diameter ($D$) is larger than the standard pin diameter ($d$), and causes the workpiece to bind or jam on the locator.

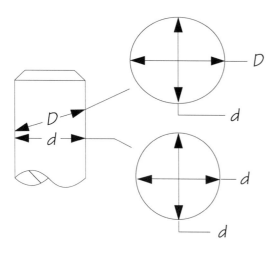

**New Method:**

No matter how a workpiece is positioned on a spherical locator, the diameter of the locator ($d$) is always the same. This feature makes the spherical locator impossible to jam or bind in the locating hole.

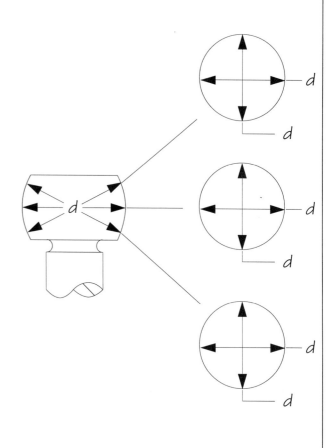

# Techniques for Locating

*Description of Problem/Requirement:*

A hole-mounted locator to minimize workpiece jamming that is also wear resistant.

*Suggested Solution:*

Design a locator with a thin contact band and relieved for easy workpiece loading and unloading.

*Source:*

In-house fabrication.

*Old Method:*

Although the spherical locator is impossible to jam or bind, the single line contact of this style locator also makes it likely to wear out rapidly. Even if the locator is hardened, the very limited contact area may cause problems if applied improperly.

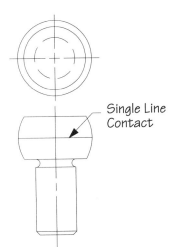

*New Method:*

To achieve both the nonjamming characteristics of the spherical locator and an extended service life for a locator, a modified form of relieved locator is necessary.

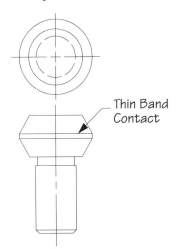

The 45° relief angles make the locator less likely to bind or jam in the workpiece, and the thin band contact design offers more resistance to wear.

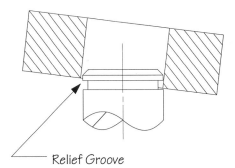

Another variation combines the thin contact band with a 45° relief angle and a relief groove.

*Description of Problem/Requirement:*

Provide a self-centering locator for cylindrical workpieces.

*Suggested Solution:*

Design a conical locator.

*Source:*

In-house fabrication.

*Old Method:*

There is no consistently reliable method of providing a self-centering location with a solid cylindrical locator. For those applications where cylindrical workpieces must be center mounted, chucks, collets, or arbors are often used. However, these may not always be appropriate for mounting workpieces in jigs or fixtures. Here, the only alternative form of solid self-centering locator is the conical locator.

*New Method:*

Internal Conical Locator

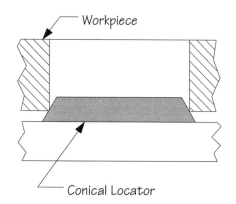

External Conical Locator

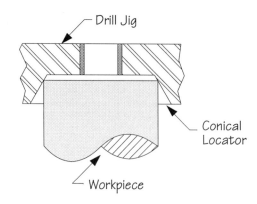

# Techniques for Locating

**Description of Problem/Requirement:**

Provide a conical locator that will locate a workpiece in three axes.

**Suggested Solution:**

Design an adjustable conical locator.

**Source:**

In-house fabrication.

**Old Method:**

The size of the mounting diameter will have an effect on the precise position of a workpiece mounted on, or in, a conical locator. With a hole mounted on a conical locator, the smaller the hole, the higher the workpiece will ride on the conical locator. If the vertical position of the workpiece is important, some type of adjustment must be built into the conical locator.

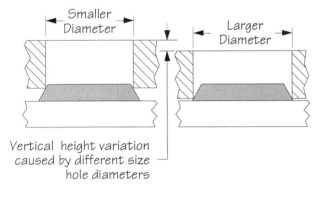

Vertical height variation caused by different size hole diameters

**New Method:**

When the vertical position of the workpiece is important, a spring-loaded adjustable conical locator may be the answer. Here the conical locator positions the workpiece along both the X- and Y-axes. The tool body establishes the workpiece position in the Z-axis. To accomplish this, a spring is mounted below the conical locator. As the workpiece is clamped, the spring is depressed and the workpiece is clamped against the locating face of the tool body. This method is not nearly as precise as a solid locator, and should only be used when sufficient locational tolerance of the workpiece is permitted.

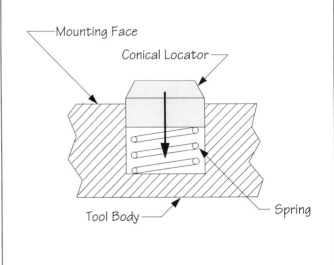

## Techniques for Locating

**Description of Problem/Requirement:**

Simplify the methods used to locate a workpiece by two holes.

**Suggested Solution:**

Use a floating locating pin arrangement.

**Source:**

Carr Lane Manufacturing Co.

**Old Method:**

Designing a locational system for mounting a workpiece by two holes involves sizing two cylindrical locators—one is a concentric locator, and the other is a radial locator. The concentric locator is made to suit the MMC (Maximum Material Condition) size of the hole. The radial locator is sized to suit the hole size tolerance of both holes as well as the locational tolerance of the holes. When making the radial locator to suit these requirements, the size of the locator is sometimes small enough to create problems in accurately locating the workpiece. Accommodating the worst case part at one end of the tolerance may cause problems when locating workpieces at the other end of the tolerance.

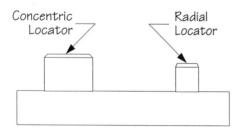

**New Method:**

The floating locating pin arrangement makes allowance for differences in the hole positions up to 1/8" in one axis while maintaining precise alignment in the opposite axis. So, while up to 1/8" movement is permitted in the X-axis direction, the pin stops any movement along the Y-axis.

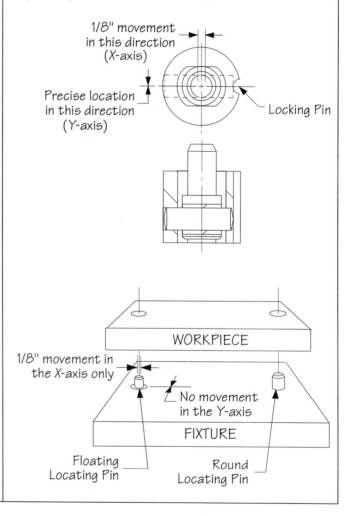

# Techniques for Locating

*Description of Problem/Requirement:*

Simplify locating cylindrical workpieces.

*Suggested Solution:*

Replace solid locators with locating pins.

*Source:*

In-house fabrication.

*Old Method:*

Locating cylindrical workpieces with solid locators adds both bulk and weight to a workholder. These locators can also add substantially to the total cost of the workholder.

*New Method:*

Replacing the solid locators with locating pins simplifies the location and reduces the total weight of the workholder. This arrangement also prevents the buildup of chips and debris, and makes the locators easier to clean.

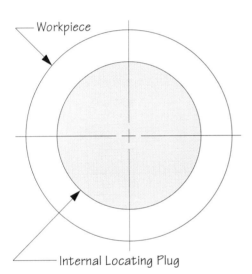

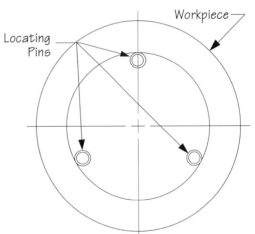

Locating on a large inside diameter with three locators spaced 120° apart.

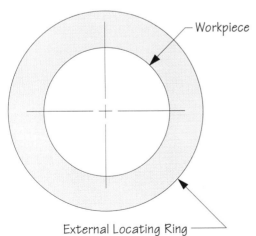

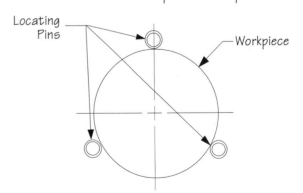

Locating on a large outside diameter with three locators spaced 120° apart.

*Description of Problem/Requirement:*

Design a simplified self-centering locator for cylindrical workpieces.

*Suggested Solution:*

Incorporate a "V"-type locator into the workholder design.

*Source:*

Various jig and fixture component manufacturers.

*Old Method:*

The "V"-type locator is an averaging locator that centrally locates the workpiece within the locator. These locating devices are well suited for a variety of simple and complex locating arrangements. This style locator may be used for any of several workpiece shapes. But, most often, they are used for round or cylindrical workpieces.

The "V"-type locators are often made in-house rather than purchased. However, to reduce the fabrication cost, when the design permits, commercial "V"-type locators should be purchased whenever practical.

The normal angle of a standard "V"-type locator is 90°. This angle may be modified when necessary to suit specific workpiece conditions. For larger diameter workpieces that must be mounted in confined areas, or where a normal 90° angle "V" locator is too large, a 120° to 140° modified "V" locator may also be used. Modifying the angle of the "V" will increase the capacity of the locator without greatly increasing its overall size.

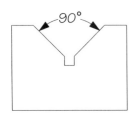
Standard "V" Locator

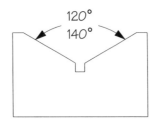
Modified "V" Locator

*New Method:*

### Standard Commercial "V" Locators

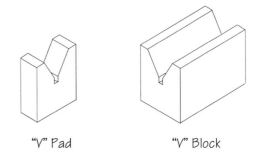

"V" Pad       "V" Block

### Typical Applications for "V" Locators

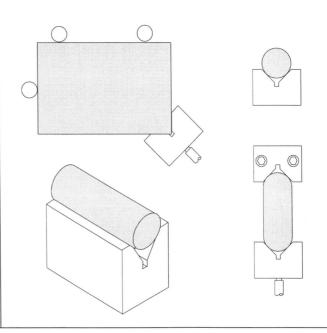

# Techniques for Locating

**Description of Problem/Requirement:**
Avoid locational inaccuracies with "V"-type locators.

**Suggested Solution:**
Properly position the "V" locator with respect to the workpiece.

**Source:**
In-house fabrication.

**Old Method:**

The averaging effect of a "V" locator can result in a locational problem if the "V" locator element is not properly referenced and positioned relative to the workpiece or the machining operation. When possible, the "V" should be positioned to minimize the effects of the workpiece size variations.

If the "V" locator is positioned on either side of the workpiece, the diameter variation, even if within allowable tolerances, will shift the workpiece. This causes misalignment of the workpiece center. This misalignment, although slight, may affect the accuracy of the complete fixturing setup.

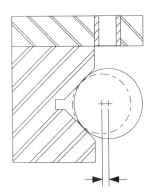

Horizontal center position error caused by variations in the workpiece diameter

**New Method:**

The alignment problem is easily overcome by repositioning the "V" locator so the "V" element is located either above or below the workpiece. Any diameter variation caused by the tolerance is taken up vertically. The size variations will then have no effect on the horizontal alignment or relative position of the workpiece.

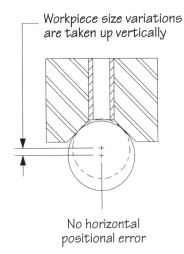

Workpiece size variations are taken up vertically

No horizontal positional error

# Techniques for Locating

**Description of Problem/Requirement:**

Design a simplified locating system that is easy to construct.

**Suggested Solution:**

Use adjustable locators in place of solid locators.

**Source:**

Various jig and fixture component manufacturers and in-house fabrication.

**Old Method:**

Solid locators are the most common style of general-purpose locator used for workholders. Although these locators are very well suited for many locating tasks, they also require considerable effort to install accurately. In most installations, the mount for these solid locators is drill and reamed. The locators are then pressed into the hole. They also require a great deal of precision to make sure they are properly positioned; and, when they wear, the complete locator must be replaced. When constructing low-cost workholders, adjustable rather than solid locators are usually a better choice.

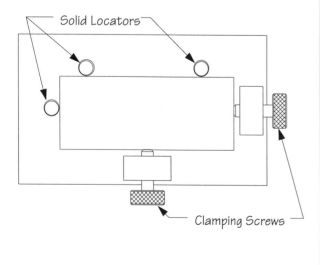

**New Method:**

Adjustable locators are usually the least expensive type of locating device. They are usually just as accurate as solid locators. However, the precision needed to mount the locators on the workholder is greatly reduced. Typically these locators are placed at the approximate position and adjusted to the exact location.

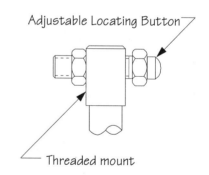

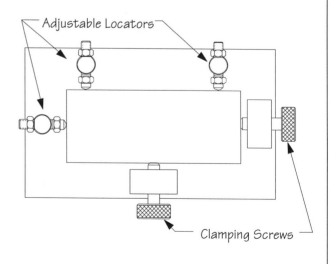

# Techniques for Locating

**Description of Problem/Requirement:**

Design more durable and less expensive fixturing supports.

**Suggested Solution:**

Replace machined supports with installed workpiece supports.

**Source:**

Various jig and fixture component manufacturers.

**Old Method:**

Often the supporting devices used for workholders are simply machined into the base element of the tool body. Although a common practice, this is also the most expensive way to provide the necessary workpiece support.

Machined supports require additional time to machine into the base; they also require the base element to be made from thicker material to allow material to machine the support. When these supports wear, they are considerably more expensive to repair.

Typically the repair process involves building up the worn support with weld, and remachining the support to the original sizes. The machined support may also be machined off, and installed supports mounted in place of the machined supports. A better approach is to design the workholder with installed supports in the beginning.

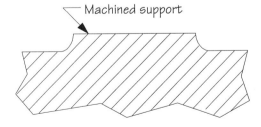

**New Method:**

## Commercially Available Supports

Solid Rest Button    Spherical Radius Locator Button    Screw Rest Button

Rest Plate    Jig Rest Button

**Description of Problem/Requirement:**

Provide an inexpensive support to accommodate different workpiece sizes.

**Suggested Solution:**

Design adjustable supports into the workholder.

**Source:**

Various jig and fixture component manufacturers.

**Old Method:**

Not all workpieces can be located on solid supports. Some workpieces, such as castings or forgings, have irregular shapes and inconsistent sizes. For this reason, some type of adjustment must be built into the workholder to allow these workpieces to be accurately located and rigidly supported.

Several methods may be used to make this adjustment, but the fastest and most common method is with adjustable locators. A simple cap screw, threaded into the tool body, may be sufficient for this adjustment, but commercially available adjustable supports offer more features and are less expensive to use in the long run. Although the most common types use a simple screw thread to make the adjustment, some styles also use a spring-loaded or hydraulic plunger.

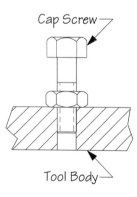

**New Method:**

<u>Commercially Available Adjustable Supports</u>

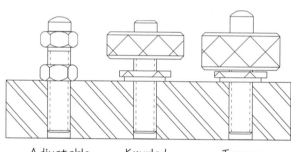

Adjustable Locating Button    Knurled Screw Jack    Torque Screw Jack

Adjustable supports are generally used along with solid supports to provide any necessary locational adjustment for the workpiece.

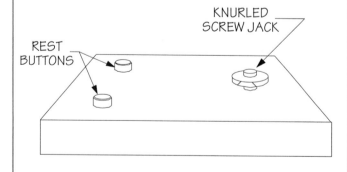

# Techniques for Locating

**Description of Problem/Requirement:**

Design a supporting arrangement to support a larger surface area.

**Suggested Solution:**

Use an equalizing-type work support.

**Source:**

In-house fabrication.

## Old Method:

Supporting some workpieces involves spreading the contact points over a wider area. This can be performed in several ways. If the surface is flat and consistent from one workpiece to the next, simply adding an additional solid support may be sufficient.

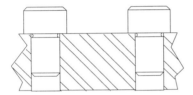

If the supporting surface is irregular, or varies considerably between workpieces, adjustable workpiece supports may also be used.

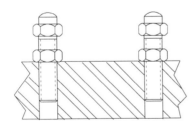

However, the additional support element required for this arrangement may cause some locational problems. Using four, rather than three, supports under a workpiece will often make accurately locating a workpiece surface difficult. If this arrangement is used, be especially careful in both selecting and positioning the supports. A better method would be to use an equalizing-type support for these workpieces.

## New Method:

An equalizing support is a self-compensating style of adjustable support. They are installed so they float to suit the workpiece surface variations. As one side moves down, the other moves up the same amount to maintain workpiece contact.

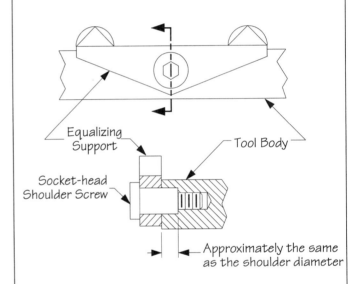

The equalizing support is typically mounted with a shoulder-type screw. To reduce any possibility of shearing the screw, the mounting hole in the tool body is counterbored. The diameter of the counterbore must closely match the diameter of the shoulder. The depth of the counterbore is approximately the same as the diameter of the shoulder.

**Description of Problem/Requirement:**

Design a supporting arrangement to support an irregular or angled surface.

**Suggested Solution:**

Use a swivel-ball-style work support.

**Source:**

Various jig and fixture component manufacturers.

**Old Method:**

Some supporting applications require a specialized work support to provide the necessary support and locational accuracy. One such application is supporting a workpiece on an angled or irregular surface. Many times, a spherical contact work support is used to locate these surfaces. Although adequate, the basic spherical form only contacts the workpiece at a single point. With some large or heavy workpieces, this could cause problems. A better solution would be to locate the workpiece on a larger size contact that also conforms to the existing workpiece surface.

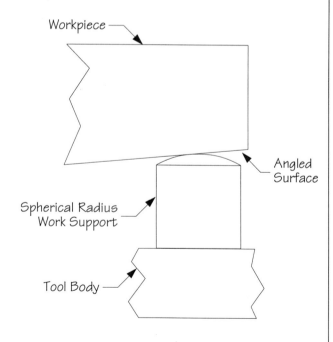

**New Method:**

The only type of work support that offers a larger contact area and also conforms to a variety of workpiece surface conditions is a swivel-ball contact. The swivel-ball contact has a flattened ball contained in a swivel mount. This work support swivels or pivots to an appropriate angle to support irregular or angular part surfaces.

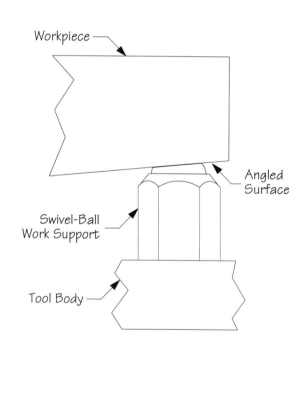

# Techniques for Locating

**Description of Problem/Requirement:**

Design a supporting system for thin or fragile workpieces.

**Suggested Solution:**

Plan a series of thrust supports into the workpiece support system.

**Source:**

In-house fabrication.

**Old Method:**

Normal locating techniques require three supports under the workpiece. However, when locating or supporting thin or fragile workpieces, the three supported points may not be properly located if the workpiece deflects during the machining operation. Here a series of additional supports must be used to absorb the thrust to minimize any workpiece deflection.

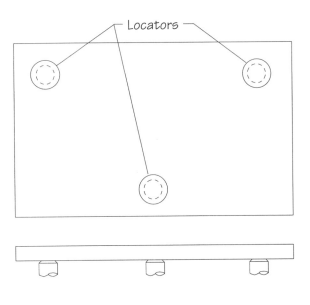

**New Method:**

Thrust supports are additional supports positioned, as needed, between the locating points. The thrust supports must not interfere with the workpiece location, so they are made between 0.0005" and 0.002" shorter than the workpiece supports. When the workpiece is loaded, it only contacts the three supports. Then, if the workpiece deflects during machining, the thrust supports will limit the amount of workpiece deflection.

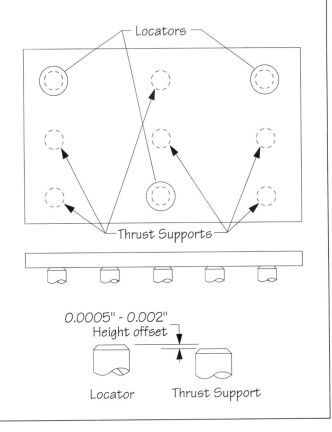

# Techniques for Locating

**Description of Problem/Requirement:**

Provide a secure method to hold the workpiece against the locators during clamping.

**Suggested Solution:**

Incorporate spring locating pins into the workholder design.

**Source:**

Carr Lane Manufacturing Co.

**Old Method:**

No matter how well a locating system is designed, unless the parts are properly positioned against the locators, mistakes result. The most common way of keeping the workpiece against the locators during clamping is by simply holding the workpiece. Some setups, however, make holding the workpiece this way difficult.

Quite often, with some arrangements, a third hand is needed to hold the workpiece while positioning and tightening the clamps. An alternative method of both positioning the workpiece while operating the clamps involves including a spring-loaded positioner in the workholder design.

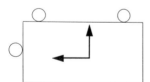

Workpiece must be held against the locators during the clamping operation.

**New Method:**

The spring locating pin pushes the part against the fixed locators. This ensures proper contact during the clamping operation. The spring locating pins also help reduce errors by correctly maintaining the workpiece position.

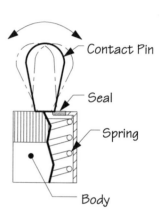

Mounting a Workpiece      Alternate Applications

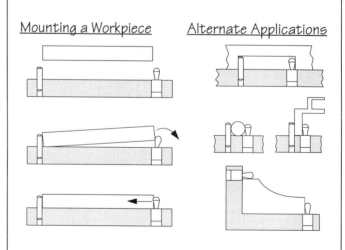

# Techniques for Locating

**Description of Problem/Requirement:**

Simplify the placement of spring locating pins.

**Suggested Solution:**

Install eccentric liners with each spring locating pin.

**Source:**

Carr Lane Manufacturing Co.

**Old Method:**

Spring locating pins are very useful for some workpiece setups. However, like fixed locators, they can also be difficult to accurately position on the workholder. Likewise, once pressed into a mounting hole, there is no way to adjust their position relative to the workpiece. An alternative to mounting these spring locating pins directly into the tool body is to mount them in an eccentric liner.

**New Method:**

The eccentric liner permits a wider locational tolerance when positioning spring locating pins. Rather than being mounted in a fixed location, these eccentric liners permit the spring location pins to be adjusted to suit the workpiece position.

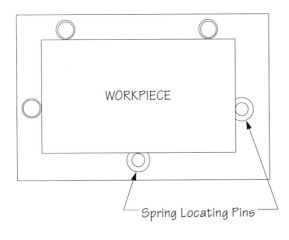

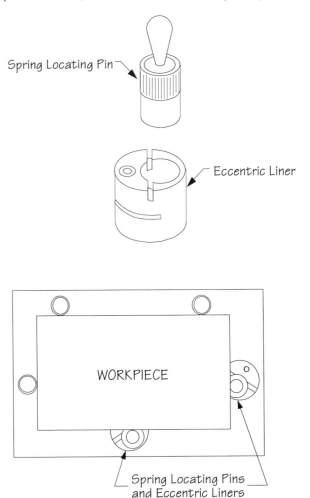

## Techniques for Locating

**Description of Problem/Requirement:**

Provide a secure method to hold larger workpieces against the locators during clamping.

**Suggested Solution:**

Incorporate spring stop buttons into the workholder design.

**Source:**

Various jig and fixture component manufacturers.

**Old Method:**

Spring locating pins are very useful for a wide range of workholding applications. However, these devices do have limitations and may not be the best choice for every workholding application. For larger workpieces, or those needing another style or shape contact surface, the spring locating pins may not be the best choice. Likewise, for those applications where a greater contact travel is required to suit different workpieces, spring stop buttons may be used in place of the spring locating pins.

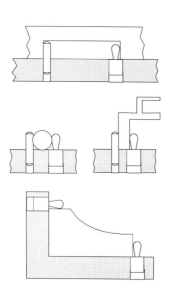

**New Method:**

Spring stop buttons are available with three styles of contacts. These spring devices are flush mounted on the tool body and held in place with screws.

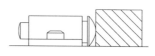

Spherical Button

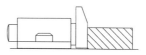

Flat Face

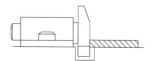

Flat Face with Tang

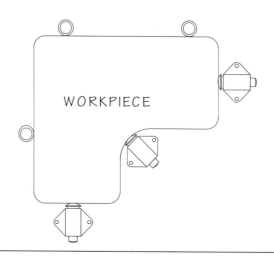

# Techniques for Locating

**Description of Problem/Requirement:**

Design a simplified locating mechanism for indexing operations.

**Suggested Solution:**

Use a retractable plunger arrangement.

**Source:**

Carr Lane Manufacturing Co.

**Old Method:**

Indexing is an operation frequently performed with some workholders. The simplest style of indexing device is a simple jig pin. This pin is mounted in the tool body and is simply pushed into an indexing hole to reference the workpiece. To advance the workpiece to the next indexing station, the pin is retracted, the fixture or workpiece is rotated, and the pin is reengaged. Although the jig pin is a common device for this purpose, other alternatives are available and should be considered when designing any indexing workholder.

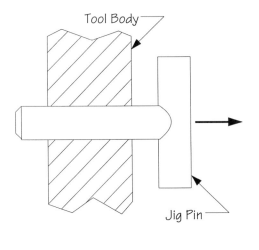

**New Method:**

Hand-retractable plungers accurately align workholder elements by engaging an indexing or referencing hole. This sets the correct position and provides a positive lock. Both plunger styles are hand operated and are pulled back to disengage the indexing hole. When the handle is released, a spring advances the plunger into the hole.

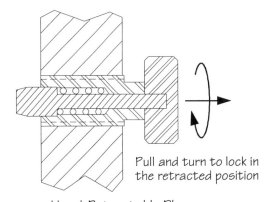

Hand-Retractable Plunger

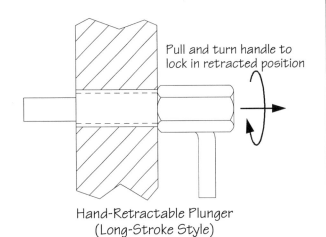

Hand-Retractable Plunger
(Long-Stroke Style)

**Description of Problem/Requirement:**

Provide a simplified automatic registration indexing mechanism.

**Suggested Solution:**

Incorporate a ball plunger assembly into the fixture design.

**Source:**

Various jig and fixture component manufacturers.

**Old Method:**

Another indexing arrangement also uses a jig pin as an indexing plunger, but engages an indexing ring rather than the workpiece. However, this arrangement also requires the indexing pin to be retracted to perform the indexing steps. A simplified variation of this indexing arrangement uses a ball plunger and detent in place of the indexing pin and indexing holes.

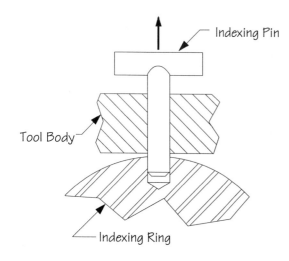

**New Method:**

Ball plungers are spring-loaded devices often used to locate workholder elements. The basic ball plunger contains a hardened ball as a plunger. In many applications, the ball plunger is combined with the ball detent. The ball detent acts as a hardened locating and referencing device for the ball plunger. To index with this arrangement, the workpiece is simply rotated until the ball plunger engages the detent.

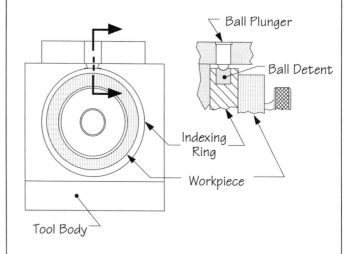

# Techniques for Locating

**Description of Problem/Requirement:**

Design an inexpensive locating system to hold complex workpiece shapes.

**Suggested Solution:**

Employ a nesting-type locator.

**Source:**

In-house fabrication.

**Old Method:**

Nesting locators are usually the most complete way to locate a workpiece. These locators can restrict up to eleven degrees of freedom. The full nest completely encases the complete periphery, or external surfaces, of a workpiece. Many times, a full nest is machined. Depending on the workpiece shape, this may be a very expensive form of locating device.

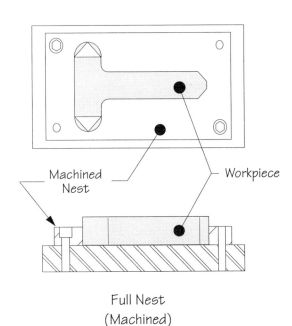

Full Nest
(Machined)

**New Method:**

One variation of the full nest that can greatly simplify the construction process is the cast nest. Cast nests are generally used for complex shapes or to hold workpieces with very irregular locating surfaces. The nest is cast to suit the three-dimensional shape of the workpiece. The most common materials used for casting full nests are an epoxy resin material or a low-melt alloy. In either case, a cast nest conforms very well to even the most intricate part shapes. In addition, since the nest is cast, the time and expense involved to make this form of nest is only a fraction of that required for a machined nest.

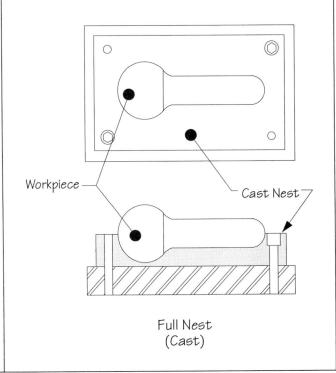

Full Nest
(Cast)

*Description of Problem/Requirement:*

Design a simplified and inexpensive form of nesting locator.

*Suggested Solution:*

Use partial nests constructed with pins or standard components.

*Source:*

In-house fabrication.

*Old Method:*

The simplest form of machined nest is the ring nest. Ring nests are actually full nests made to suit a round or cylindrical workpiece shape. Since the basic shape of this style nesting locator is relatively simple, ring nests are usually machined rather than cast. In most cases, the same design rules and guidelines that control the size of locating holes also determine the sizes of ring nests.

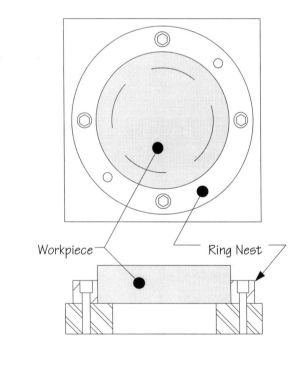

*New Method:*

A variation of the full nest is the partial nest. A partial nest acts in the same way as a full nest; however, rather than enclosing the complete external surface of the workpiece, only selected areas are located. This style of nest performs the same functions as a full nest but is much easier to make. These nests may be machined, cast, or made up of standard fixturing components.

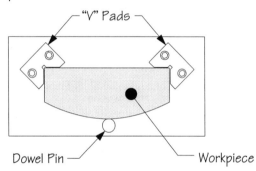

The dowel pin nest is another style of partial nest that completely locates a workpiece. Here, rather than machined elements, dowel pins completely restrict the workpiece. The dowel pin nest is one of the simplest and least expensive forms of nesting locators.

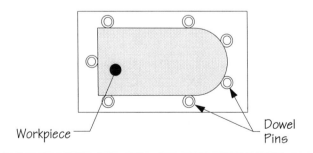

# Techniques for Locating

**Description of Problem/Requirement:**

Design a device to prevent the workpiece from being loaded upside down in the workholder.

**Suggested Solution:**

Incorporate a foolproofing device into the workholder design.

**Source:**

In-house fabrication.

**Old Method:**

Many times, the basic shape of a workpiece may allow it to fit into a workholder in several different ways. Almost always, however, there is only one correct position for the workpiece. Preventing the workpiece from being mounted incorrectly is a primary function of foolproofing. One frequent loading problem is where the workpiece can be mounted upside down. To prevent improper loading, a foolproofing device should be included in the workholder.

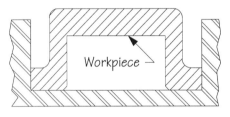

Correctly Loaded

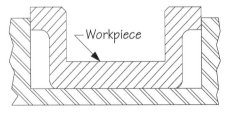

Incorrectly Loaded

**New Method:**

The specific shape or form of the workpiece will determine the type of foolproofing devices used. However, if the workpiece has either a hollow recess or a hole, this feature can easily foolproof the workpiece location. The simplest foolproofing device for this workpiece is either a block or set of pins. These will fit into the recess or hole when the workpiece is mounted correctly. The block or pins will also prevent the workpiece from being loaded if it is flipped over. The foolproofing device is not a locator and should not contact the correctly installed workpiece.

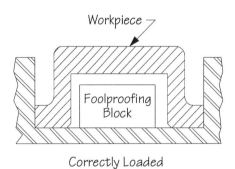

Correctly Loaded

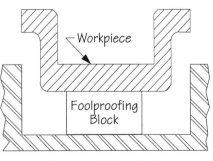

Incorrectly Loaded

| Description of Problem/Requirement: |
|---|
| Design a device to prevent the workpiece from being incorrectly loaded in the workholder. |

| Suggested Solution: |
|---|
| Incorporate a foolproofing device into the workholder design. |

| Source: |
|---|
| In-house fabrication. |

*Old Method:*

Radial workpiece locators are often used to set the radial position of a workpiece with respect to a center axis. These locators may also prevent the workpiece from being loaded incorrectly. Generally, when a radial locator is used for this purpose, the workpiece may be located in any of several positions and still be correct. However, in those cases where the radial locator may be placed in more than one location, but the workpiece only has one correct position, a foolproofing device should be used along with the radial locator.

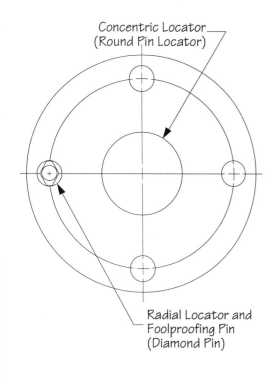

*New Method:*

When there is only one correct workpiece position, and a concentric and radial locator combination cannot guarantee this position, a foolproofing device is added. As shown here, the workpiece can be positioned in any of three locations using only the concentric and radial locators. However, due to the flanged area of the workpiece, only one of these three positions is correct. Therefore, a foolproofing pin is positioned so that the workpiece can only be loaded in the correct position.

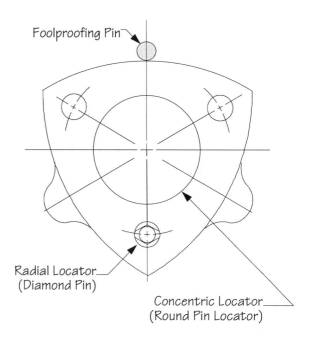

# Techniques for Locating

**Description of Problem/Requirement:**

Design a locating method to prevent incorrect loading of the workpiece.

**Suggested Solution:**

Design a specialized radial locator to foolproof the location.

**Source:**

In-house fabrication.

**Old Method:**

With some workpieces, the only features available for location are a center hole and an external feature. The center hole can be used for concentric location, and the external feature for radial location. A pin, positioned against the external feature, will radially locate the workpiece. But, in some cases, the workpiece may be positioned on the wrong side of the radial locating pin. To prevent this incorrect loading, a modified radial locator can also serve as a foolproofing device.

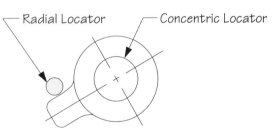

Loaded Correctly

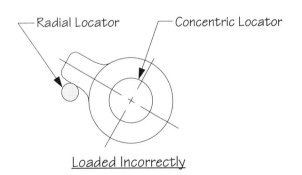

Loaded Incorrectly

**New Method:**

To prevent the part from being loaded on the wrong side of the radial locating pin, either two pins or a single locating block can be used as a radial locator. When two pins are used, the distance between the pins should be controlled to allow the workpiece to be loaded correctly when contacting either pin. If a block is used, the shape of the block should make the correct location obvious.

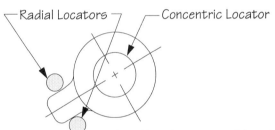

Locating pins are positioned so the location is correct when contacting either pin.

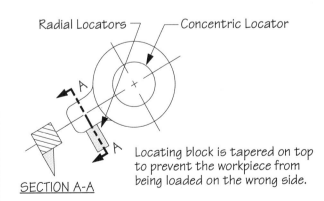

SECTION A-A

Locating block is tapered on top to prevent the workpiece from being loaded on the wrong side.

# CHAPTER FIVE

*Techniques for Clamping*

# Techniques for Clamping

**Description of Problem/Requirement:**
Determine the best type of strap clamp arrangement for a clamping application.

**Suggested Solution:**
Select between a first-, second-, or third-class lever arrangement for the strap clamp.

**Source:**
Various jig and fixture component manufacturers and in-house fabrication.

**Old Method:**

The simplest and most common form of clamp used today is the strap clamp. This clamp consists of a flat clamp strap, a stud or bolt assembly, and a heel support, or heel pin. While there are variations of this simple design, most have these three elements.

All strap clamps work on the mechanical principle of the lever. In use, the clamps hold the workpiece by the force generated by the fastener, transferred through the clamp strap, to the workpiece. The heel pin acts as a pivot and supports the back end of the clamp strap.

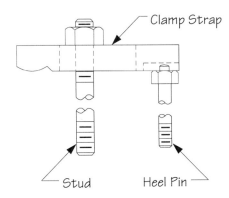

**New Method:**

The three basic styles of strap clamps are described in terms of first-, second-, or third-class levers. These classes of levers do not indicate importance or preference, but distinctions in the basic mechanical actions. The difference in these three variations is in the placement of the effort, the part, and the fulcrum.

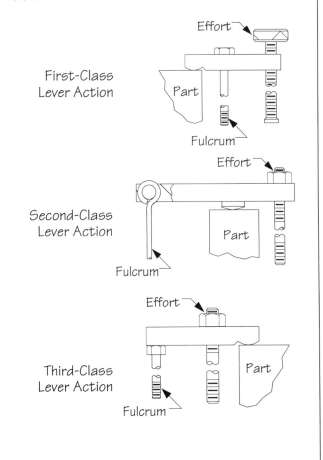

**Description of Problem/Requirement:**
Design an economical clamp strap to suit the workpiece parameters.

**Suggested Solution:**
Select a commercial clamp strap suitable for the workpiece being clamped.

**Source:**
Various jig and fixture component manufacturers.

**Old Method:**

Clamp straps are basically a flat piece of metal with a hole drilled to accommodate a stud, bolt, or other threaded fastener. Almost any flat bar will work as a clamp strap. However, to achieve the desired clamping action, specific types or styles of clamp straps are designed for different applications. The clamp strap should be selected to suit the specific conditions of the workpiece and the clamping operation.

Just as the style of clamp strap is important, so too is the style of the clamping contact. Here, again, the clamp contact should be selected to suit the workpiece requirements.

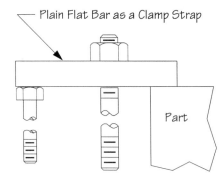
Plain Flat Bar as a Clamp Strap
Part

**New Method:**

<u>Clamp Strap Variations</u>

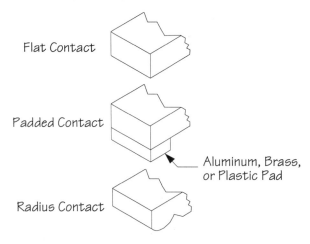

Plain Strap
Tapered Nose
Wide Nose
"U"-Shaped
Gooseneck
Double End

<u>Clamp Strap Clamping Contacts</u>

Flat Contact

Padded Contact

Aluminum, Brass, or Plastic Pad

Radius Contact

# Techniques for Clamping

**Description of Problem/Requirement:**
Design an economical heel supporting device to suit the workpiece parameters.

**Suggested Solution:**
Select a commercial heel support arrangement suitable for the workpiece being clamped.

**Source:**
Various jig and fixture component manufacturers.

## Old Method:

The heel support is located at the end of a clamp strap, opposite the contact area. The heel support acts as a pivot and also supports the back end of the strap. Two of the more common methods of supporting the heel of a clamp strap are with stacked pieces of scrap stock or with a step block arrangement. However, these are not the only types of heel supports. Heel supports are available in several styles to suit the wise use of strap clamps. In most cases, the type of clamp strap used as well as the application of the clamp will decide which style of heel support is best.

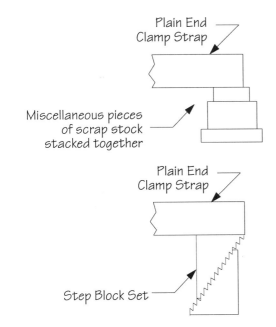

## New Method:

### Heel Support Variations

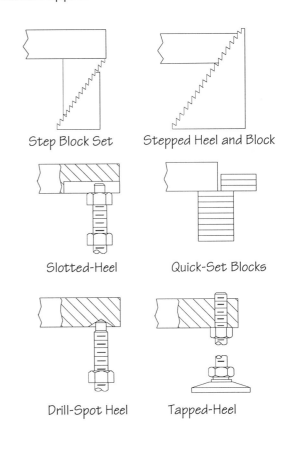

Step Block Set — Stepped Heel and Block

Slotted-Heel — Quick-Set Blocks

Drill-Spot Heel — Tapped-Heel

*Description of Problem/Requirement:*
   Design an efficient method of minimizing the effects of fatigue on the threaded fasteners.

*Suggested Solution:*
   Select a spherical nut and washer or spherical washer set.

*Source:*
   Various jig and fixture component manufacturers.

*Old Method:*

Fatigue occurs from the constant stress applied on threaded fasteners. This is caused by repeatedly clamping workpieces of varying heights. All workpieces within a production run are made to variable sizes, within their stated tolerances. Even slight differences can cause fatigue, over time, that weakens any clamping stud or bolt.

Some of the more common styles of fastening devices used for strap clamps are the standard nut/washer arrangement or the flange nut. Although widely used, neither of these nuts can compensate for the different workpiece heights. A better choice would be a spherical nut and washer set or a spherical washer assembly.

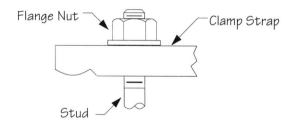

*New Method:*

The spherical form acts as a universal joint between the clamp and the stud or bolt. This arrangement allows limited angular movement of the clamp strap with no effect on the fastener. The spherical joint eliminates the stress on the stud or bolt because it compensates for the angular misalignment of the clamp strap and the fastener.

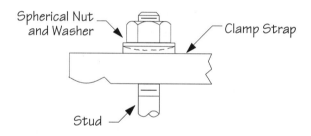

Spherical washer sets are available in two styles: a spherical washer and nut combination, and a set of washers with mating spherical faces. The spherical nut and washer unit is best applied to new clamping setups, while the washer sets are often used to retrofit older clamps.

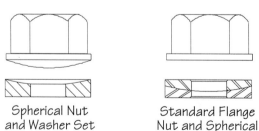

# Techniques for Clamping

*Description of Problem/Requirement:*
Design a strap clamp arrangement to achieve the best holding forces.

*Suggested Solution:*
Properly position the stud with respect to the workpiece and heel support.

*Source:*
In-house fabrication.

## Old Method:

To achieve the maximum mechanical advantage when applying a strap clamp, the position of the clamping stud with respect to the workpiece and heel support is important. If the stud is positioned exactly in the center of the clamp strap, the pressure generated by the fastener is distributed equally between the workpiece and the heel support. Moving the stud closer to the workpiece increases the mechanical advantage and the holding force.

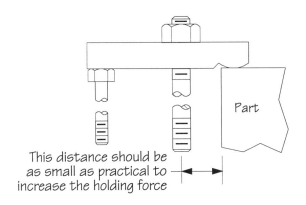

This distance should be as small as practical to increase the holding force

## New Method:

With the stud in the center of the clamp strap, equal force is applied to both the workpiece and the heel support.

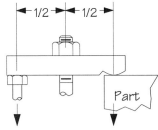

1/2 Clamp Force    1/2 Clamp Force

This arrangement is well suited for applications where two workpieces are clamped.

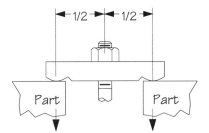

1/2 Clamp Force    1/2 Clamp Force

Where a single workpiece is clamped, moving the stud toward the workpiece increases the mechanical advantage of the clamping.

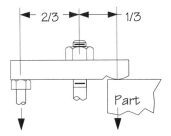

1/3 Clamp Force    2/3 Clamp Force

## Techniques for Clamping

**Description of Problem/Requirement:**
Design a clamping arrangement to minimize the effect of the clamp height.

**Suggested Solution:**
Select the correct clamp strap and fastener length to suit the workpiece requirements.

**Source:**
Various jig and fixture component manufacturers and in-house fabrication.

### Old Method:

The overall height of the clamp is a problem that arises with some stud and nut arrangements. The general rule is to always keep the height of the clamping elements as low as possible. In those cases where the height of a stud and nut could be objectionable, such as workholders used close to arbors or spindles, a bolt might be a better choice. The height of a bolt head is considerably less than the height of a comparable stud and nut arrangement. This allows a strap clamp to be used in areas where space is limited.

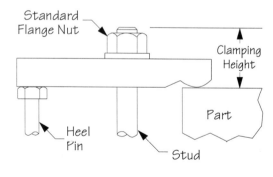

### New Method:

A bolt may be used in place of a stud and nut arrangement where the total height of the clamping assembly is important.

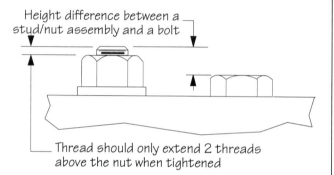

Where a standard plain strap clamp assembly extends too far above the workpiece surface, a gooseneck-style strap clamp may be used to reduce the overall height of the clamp assembly.

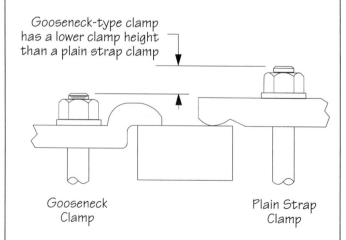

# Techniques for Clamping

**Description of Problem/Requirement:**
Provide a strap clamp suitable for applications where space is limited.

**Suggested Solution:**
Incorporate an adjustable-style strap clamp into the workholder design.

**Source:**
Various jig and fixture component manufacturers.

**Old Method:**

Strap clamps, although used for a wide range of varied applications, are not well suited for every workholding task. There are times when the overall size or mass of a standard strap clamp can be a problem. Here, the clamping requirements are often the same, but the area for the clamp is limited. Occasionally, simply eliminating the heel support may reduce the overall size enough to allow the clamp to fit into a tight space.

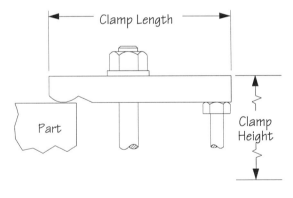

**New Method:**

The forged adjustable clamp is well suited for places where a heel support can interfere with the setup. These clamps are complete units and can accommodate a variety of workpiece clamping heights. A steel pivot allows the fastening element to securely hold the clamp at all elevations. Another variation of this style of clamp is a one-piece forging. With this design, a curved top surface of the clamp strap, rather than a steel pivot, positions the fastening element.

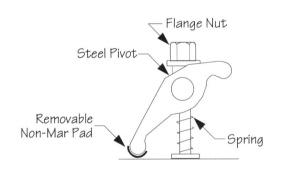

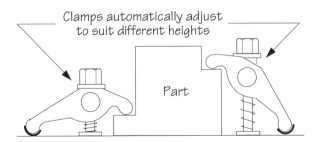

*Description of Problem/Requirement:*

Design a strap clamp arrangement for extended height clamping situations.

*Suggested Solution:*

Incorporate high-rise clamps into the workholder design.

*Source:*

Various jig and fixture component manufacturers.

*Old Method:*

Some setups involving strap clamps require a rather extended clamping length. Although a standard strap clamp can handle the extended reach, the higher the strap clamp arrangement, the less stable the entire setup becomes. Any setup over approximately 12" high becomes questionable. A better alternative is using a high-rise strap clamp assembly. These units are both more stable and offer more clamping versatility. The high-rise unit is made up of several different components that may be positioned as required to suit the workpiece.

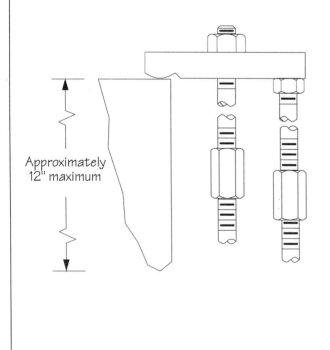

*New Method:*

High-rise clamps are assembled from a variety of different elements. Depending on the specific manufacturer, a range of clamp straps, contacts, and several other accessories are also available.

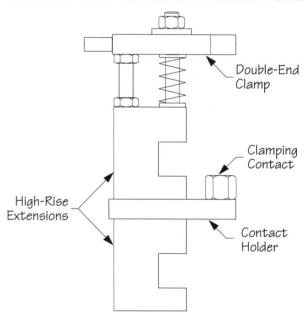

<u>Typical Setup for a High-Rise Clamp Assembly</u>

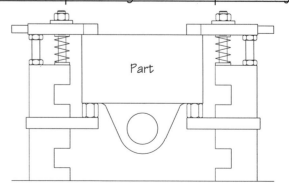

# Techniques for Clamping

**Description of Problem/Requirement:**
Design a strap clamp arrangement that prevents workpiece distortion during clamping.

**Suggested Solution:**
Replace standard strap clamps with S.A.F.E. clamping elements.

**Source:**
Safe-Tech Corporation.

**Old Method:**

When irregular or odd shape workpieces are clamped with conventional strap clamps, the workpiece will sometimes deform or distort. When unclamped, the workpiece returns to its original shape. Therefore, all machining performed when the workpiece was clamped is also distorted and not as originally machined.

## Conventional Strap Clamps

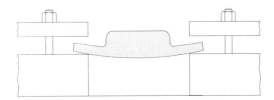

Workpiece distorts as the clamps are tightened

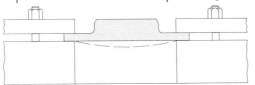

When the clamps are released, the workpiece returns to its original shape and the machined surface is not flat.

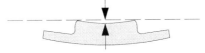

**New Method:**

The S.A.F.E. ball elements permit the contact surface of the locators, supports, and clamps to conform to the workpiece shape. This eliminates almost all chance of part distortion during clamping.

## S.A.F.E. Style Clamps

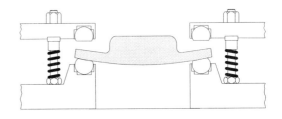

The ball elements conform to the irregular workpiece shape to prevent distortion when clamped.

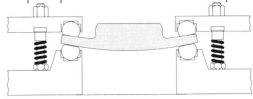

When the clamps are released, the workpiece remains flat, since there was no workpiece distortion.

## Techniques for Clamping

**Description of Problem/Requirement:**
Design a fast-acting screw clamp arrangement.

**Suggested Solution:**
Integrate swing-type clamps or hook clamps into the workholder design.

**Source:**
Various jig and fixture component manufacturers.

**Old Method:**

Screw clamps are one of the simplest forms of clamping devices. Their clamping efficiency, effectiveness, and size/force ratio are very good. However, the slow actuation speed inherent in the screw threads has blocked their widespread use in many workholders.

The basic mechanical principle behind the screw thread is the inclined plane. When applied to a cylindrical feature, this inclined plane becomes a helix. If the thread was unwrapped from a screw, the resulting form of the thread, relative to the screw, would be triangular. Each part of the triangle has a specific relationship to the screw. The hypotenuse represents the actual length of the thread around the screw. The side adjacent reproduces the circumference of the screw, and the side opposite shows the lead (or the amount of lateral movement the thread makes along the screw in one revolution).

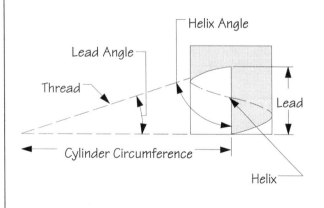

**New Method:**

One way to increase the speed of a threaded clamp is by adding a moving arm. Swing clamps are basically screw clamps with a swinging clamp arm to speed clamping and unclamping. A smaller variation of the swing clamp principle is the hook clamp. Like swing clamps, hook clamps come in a variety of sizes and styles.

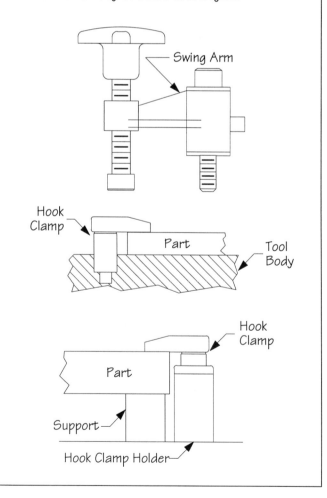

# Techniques for Clamping

*Description of Problem/Requirement:*
   Design a clamping arrangement with a fast-acting screw-type clamping device.

*Suggested Solution:*
   Design the clamping device to include a quick-acting screw or knob.

*Source:*
   Various jig and fixture component manufacturers.

**Old Method:**

Although the screw thread is a very secure form of clamping device, the time required to operate the screw is often objectionable. For this reason, the basic screw thread is not used for workholding as much as it should be. However, simply modifying the way the thread is applied in the workholder can often achieve higher clamping and unclamping speeds.

The lateral movement of a thread depends on the pitch of the thread. The pitch is the distance between two adjacent threads. The pitch is also the amount of lateral movement a thread makes in one revolution of the screw.

Lateral Movement = Revolutions × Pitch

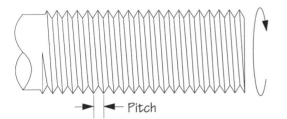

Quick-acting screws and knobs are another way to increase the speed of a threaded clamp. Several variations of quick-acting clamps are available. Each of these designs increases the speed of the screw clamp by removing a portion of the threads. Quick-acting clamp screws have a portion of the external thread removed. Quick-acting clamp knobs have a portion of the internal threads removed.

**New Method:**

<u>Quick-Acting Screw</u>

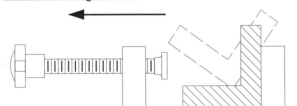

Slides back quickly for faster loading/unloading.

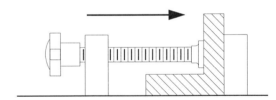

Slides forward quickly for faster workpiece contact.

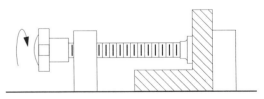

Full clamping force with less than one-turn of the screw.

<u>Quick-Acting Knob</u>

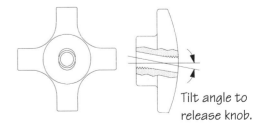

Tilt angle to release knob.

**Description of Problem/Requirement:**
Design a clamping arrangement for clamping on tabs.

**Suggested Solution:**
Incorporate a ball element into the clamp to minimize distortion when clamping.

**Source:**
Safe-Tech Corporation.

**Old Method:**

Tabs are often designed into some workpieces to make locating and clamping easier. These tabs are then drilled to suit a threaded fastener. The fasteners used for these applications vary to suit the requirement of the workpiece, but a hex head screw or a socket head cap screw are two common fasteners for these applications. Although many workpieces can be mounted directly to a tool body or custom made riser element, if there are irregularities in the workpiece, distortion could result if the tabs are not properly supported.

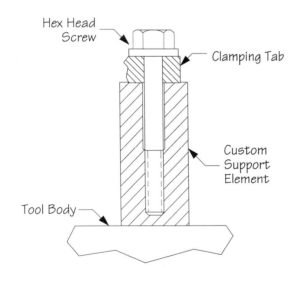

**New Method:**

The ball element-type fixture units are unique clamp variations intended to hold the workpiece with a threaded fastener rather than a clamping bar. The ball element in these units has a precision clearance hole through the ball to allow the screw to pass through the ball to attach to the clamp body. This arrangement provides ample clamping force, and the ball elements minimize any distortion when clamping.

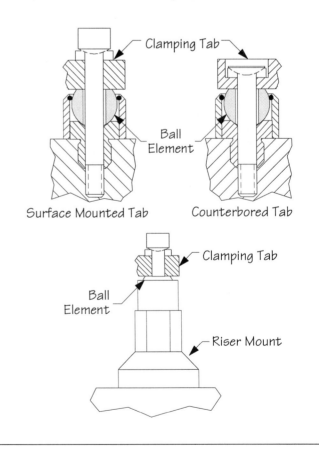

# Techniques for Clamping

**Description of Problem/Requirement:**
Design an edge clamping arrangement for larger workpieces.

**Suggested Solution:**
Add a set of pivoting-style edge gripping clamps to the workholder design.

**Source:**
Carr Lane Manufacturing Company and Safe-Tech Corporation.

## Old Method:

Edge clamping is an operation frequently performed with a variety of workholders. Many times, standard machine vises are used for these operations. However, there are occasions where these vises do not have the capacity necessary to hold some workpieces. In these cases, there is a wide range of standard and specialty edge gripping clamps and clamp sets to choose from.

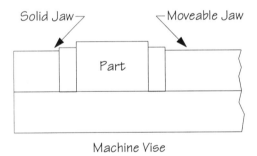

Machine Vise

## New Method:

The pivoting-style edge clamps use a pivoting clamping arm to generate the required holding force. These clamps are generally used with a matching backstop unit. Depending on the manufacturer, the back stop unit may be a fixed locator, or a type that automatically pulls the workpiece down as the clamp is activated.

Carr Lane Pivoting Edge Clamp and Backstop

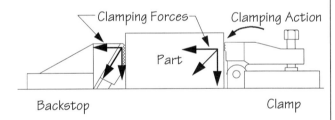

S.A.F.E. Horizontal Clamping Unit and Positioner

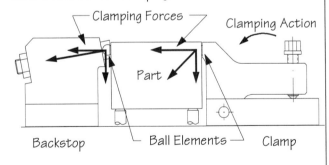

*Description of Problem/Requirement:*

Design a low-profile edge clamping arrangement for thin workpieces.

*Suggested Solution:*

Incorporate the flat-type or slot-type edge clamps into the workholder design.

*Source:*

James Morton Inc.

*Old Method:*

Edge clamping thin workpieces is a job where the size of the clamp is very important. Many times, to maintain a lower profile, either machine vises or custom made screw clamp arrangements are used. Here, the screw clamps are usually made in-house with an assortment of commercial screws. These are typically installed in special mounting blocks or other devices. Although adequate, the cost of these units is often very high. Likewise, the limited contact area offered by a clamp screw, directly contacting the workpiece, limits the amount of applied clamping force.

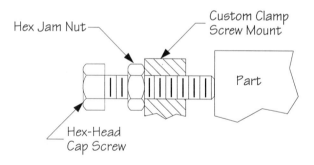

*New Method:*

An alternative to this method is using low-profile edge clamps. Two designs well suited for a wide range of clamping applications are the flat clamps and slot clamps. These clamps have a clamping action that applies forces in both horizontal and downward directions.

*New Method, continued:*

The flat clamp is intended for low-profile clamping of flat workpieces. The clamp is mounted on the machine table and operates by tightening the screw in the clamp element. This forces the clamping element against the workpiece. A spring connector holds the clamp element off the table.

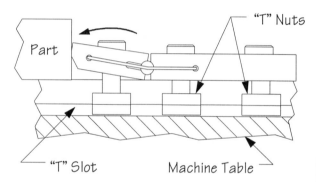

The slot clamp also applies horizontal and downward clamping forces. Slot clamps are designed to hold very thin workpieces and are mounted directly in the "T" slot rather than on the machine table. This low-profile design allows the clamps to hold workpieces as thin as 0.13" without distortion.

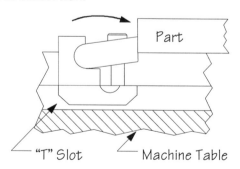

# Techniques for Clamping

*Description of Problem/Requirement:*
    Design a clamping system to mount multiple workpieces on a pallet system.

*Suggested Solution:*
    Design the workholding arrangement around the Vector or Mitee-Bite Clamping Systems.

*Source:*
    SMW Systems Inc. and Mitee-Bite Products Inc.

*Old Method:*

Pallets and pallet systems are rapidly becoming a standard part of manufacturing. But, too often, their potential benefits are not realized. Unfortunately, many pallet systems purchased today are not equipped with a compatible workholding system. Many times, these pallet systems are treated as merely supplementary machine tables and used as bases for existing fixtures or vises. When multiple workpiece setups are made, they often involve rather simplistic workholding devices. These devices, such as a simple bar with tapped holes for screws, although useful, are not nearly as universal as the pallet systems. To achieve the maximum benefit from any pallet system, the workholding elements should be as universal as the pallets themselves.

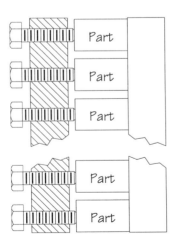

*New Method:*

Two systems that maximize the design flexibility and universality of a pallet system are the Vector Clamping System and the Mitee-Bite Clamping System.

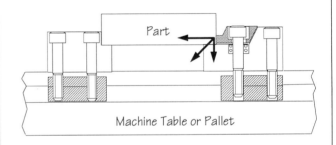

Both systems use a special clamping element along with a locator or backstop arrangement. They also raise the workpiece off the pallet to assure accurate location. The Vector System uses a clamp jaw mounted on an angular ramp. As the screw is tightened, the jaw both pulls the workpiece down and clamps it against the backstop. The Mitee-Bite uses a cam locking screw to perform the clamping.

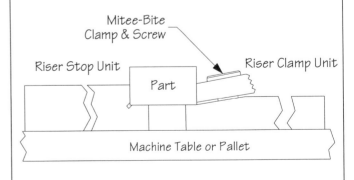

## Techniques for Clamping

**Description of Problem/Requirement:**
Design a clamping system for mounting multiple small workpieces on a pallet.

**Suggested Solution:**
Use a workholding arrangement with the Mitee-Bite fixture clamps.

**Source:**
Mitee-Bite Products Inc.

**Old Method:**

When pallets are used for multiple workpiece setups, it is usually desirable to mount as many workpieces as possible in a single setup. This is more efficient and economical than machining the workpieces one at a time. To achieve the maximum workpiece density on a pallet, the clamps used must be quite small. Simple screw clamp arrangements are often used for these setups. However, although they work well for the workpieces around the outer edges of the pallet, due to the added space requirements to operate the clamps, screw clamps may not be as efficient for the workpieces in the center.

**New Method:**

An alternative clamping method is using the Mitee-Bite Clamping System. These clamps are operated from the top, and may be packed quite closely together on the pallet.

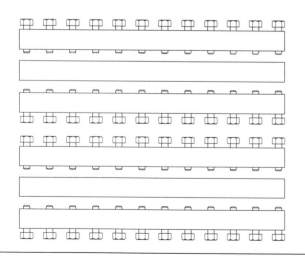

**New Method, continued:**

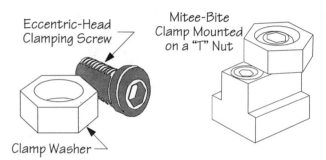

The Mitee-Bite either advances or retracts the washer when the eccentric-head screw is turned.

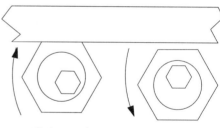

The small size of the Mitee-Bite clamps allows more workpieces to be clamped on a pallet.

# Techniques for Clamping 99

**Description of Problem/Requirement:**
Design a clamping system suitable for high-density clamping for pallet setups.

**Suggested Solution:**
Use a workholding arrangement with the Mitee-Bite Uniforce clamps.

**Source:**
Mitee-Bite Products Inc.

## Old Method:

Reducing the size of the clamping devices is an important consideration when clamping multiple parts on a subplate or pallet. Threaded fasteners are typically the most common type of clamping device made in-house. These arrangements work very well and are also very economical. However, they do lack the compactness necessary for high-density workpiece loading. Too much space is required for the screws. Even when socket-head set screws are used, there must still be enough space to use the wrench.

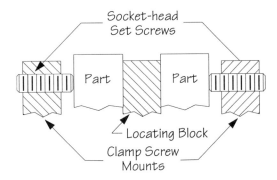

## New Method:

The Uniforce clamp is an alternative to using set screws. These clamps are operated from their top surface and do not require any additional space for a wrench. The clamp also exerts a clamping force in two directions, to clamp two workpieces at the same time.

## New Method, continued:

### Uniforce Clamp

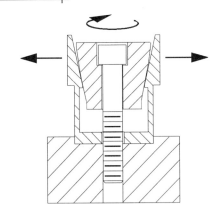

As the clamp screw is tightened, the clamp expands to apply equal holding force on two workpieces.

### Multi-Force Workholding System

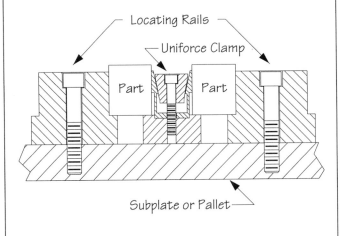

*Description of Problem/Requirement:*
  Design a clamping system for large or odd-shaped workpieces.

*Suggested Solution:*
  Specify edge-type clamps to hold the workpiece.

*Source:*
  Various jig and fixture component manufacturers.

*Old Method:*

Large or odd-shaped workpieces are often very difficult to hold on machine tables or subplates, especially if the top surface of the workpiece must be machined. Both the size and irregular contours or features can make securely holding the workpiece quite complicated. A wide range of standard clamps or vise arrangements may be used for these setups. However, a less complicated and very secure method of holding these workpieces is with simple edge, or toe, clamps. Unlike many other styles of clamps, these edge clamps may be positioned almost anywhere, in any position, to suit the workpiece requirements.

*New Method:*

Edge clamps are made in a variety of styles. However, each variation operates by transferring the rotary motion of the clamping screw into a linear clamping motion that moves both forward and downward along the clamping angle. These clamps are also made with an angular mounting slot. The 1° angle in the bottom of the slot prevents movement of the clamp body away from the workpiece. These clamps are made with either a low nose or high nose design. Most styles have gripping serrations on the clamping jaws. But many are also available with an aluminum cap to prevent damaging or marring the workpiece.

*New Method, continued:*

### Edge Clamps

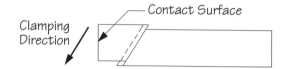

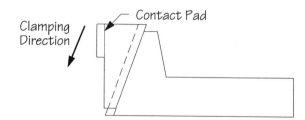

Edge clamps apply horizontal and vertical clamping force.

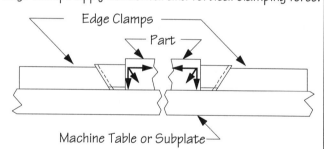

Edge clamps may be positioned wherever required.

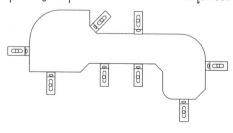

# Techniques for Clamping

**Description of Problem/Requirement:**
Design a universal clamping system for edge clamping with screw clamps.

**Suggested Solution:**
Add screw-type edge clamps to the workholder design.

**Source:**
Various jig and fixture component manufacturers.

## Old Method:

The basic screw thread is one of the most common devices employed for clamping in a wide range of jigs and fixtures. Typically these clamps are made in-house by simply drilling and tapping holes in one, or more, of the fixturing elements. Although adequate, this method is expensive and lacks the universality required for many clamping situations.

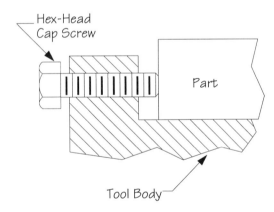

## New Method:

An alternative to machining the tool body to suit a screw clamp is the screw-type edge clamps. These clamps are workholding devices that combine a screw clamp and locator into a single unit. The screw-type edge clamps are available as either fixed or adjustable locating heights. Matching edge supports are also available, as are riser elements for elevated applications. The adjustable height units may be fitted with any of several different types of contact elements.

## New Method, continued:

### Screw-Type Edge Clamps and Fixed Supports

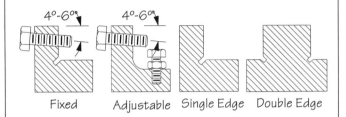

The adjustable screw-type edge clamp will usually have a variety of accessory workpiece contacts.

Riser blocks are used to elevate the clamps and supports for larger workpieces.

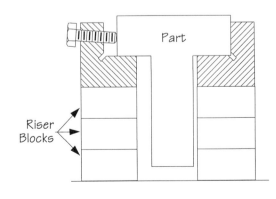

*Description of Problem/Requirement:*

Develop a clamp that is faster and capable of clamping a workpiece in two directions.

*Suggested Solution:*

Incorporate a two-directional edge clamping bar into the workholder design.

*Source:*

In-house fabrication.

*Old Method:*

The workpiece is clamped with two separate socket-head cap screws. This design requires twice the clamping time since the operator must tighten and loosen two individual clamping screws. This design can also cause uneven clamping pressure if the screws are not tightened exactly the same.

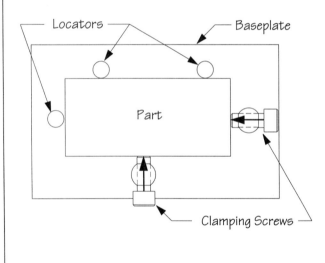

*New Method:*

A custom-made two-directional edge clamping bar allows the workpiece to be clamped in two directions at the same time, with a single clamping step. Since a single screw is tightened and loosened, the clamping time is reduced and the clamping pressure is equalized in both directions on the workpiece.

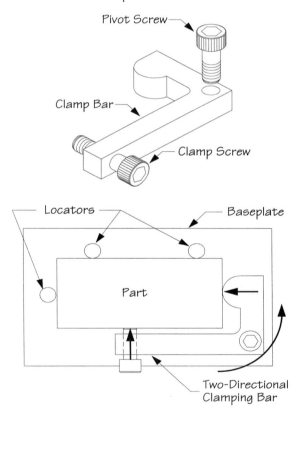

# Techniques for Clamping

**Description of Problem/Requirement:**
Design a fast acting clamping system with a small footprint to replace strap clamps.

**Suggested Solution:**
Specify Mono-Block clamps for the workholder.

**Source:**
Royal Products.

## Old Method:

Strap clamps, for all their many benefits, do have a few drawbacks for some workholding applications. Most strap clamp arrangements are time consuming to set up and operate. The threaded fasteners used to apply the clamping force require more time to operate than other types of mechanical devices.

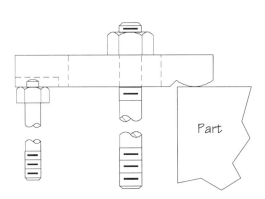

## New Method:

A fast-acting and economical alternative to standard strap clamps is the Mono-Bloc Clamping System. These clamps operate on a worm and worm wheel principle which is considerably faster than any clamping system that uses a screw thread. The compact size of these clamps also allows them to be positioned in areas where space is limited.

## New Method, continued:

### Mono-Bloc Clamp

The Mono-Bloc clamp uses a worm to drive a worm wheel to raise and lower the clamping arm. The swivel contact allows the clamp to align itself to any clamp height.

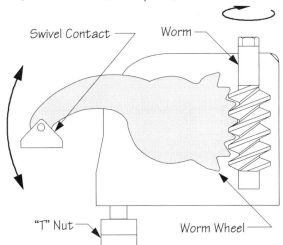

There are two standard arms for the Standard-Duty-type Mono-Bloc clamp. Depending on the size of the arm, the overall clamping height capacity ranges from 0" to 4" with a clamping throat of 1.38" to 2.38".

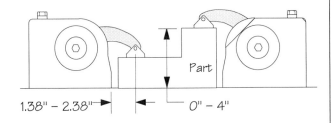

*Description of Problem/Requirement:*

Design a fast-acting clamping system with a small footprint to replace strap clamps.

*Suggested Solution:*

Specify Mono-Bloc clamps for the workholder.

*Source:*

Royal Products.

*Old Method:*

Strap clamps, when used for clamping very large workpieces, can become quite unwieldy and difficult to handle. Higher clamping heights normally require both longer studs and heel pins. This added length can make the complete setup wobbly and unmanageable. Likewise, the added height makes the clamping operation more difficult since the clamp strap is harder to position properly.

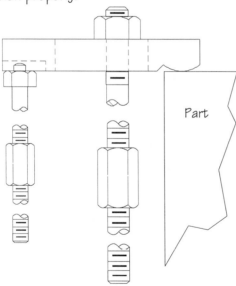

*New Method:*

The Mono-Bloc clamping system, however, is very easy to set up for larger workpieces. Simply adding an extension arm and riser block will extend the range of the clamps with no loss in the overall stability of the assembly.

*New Method, continued:*

### Mono-Bloc Clamp Accessories

The Mono-Bloc clamps can be configured for larger workpieces by simply adding an extension arm or riser blocks. These elements permit the range of the clamp to be extended to 12" high with a maximum clamping throat of 5.63".

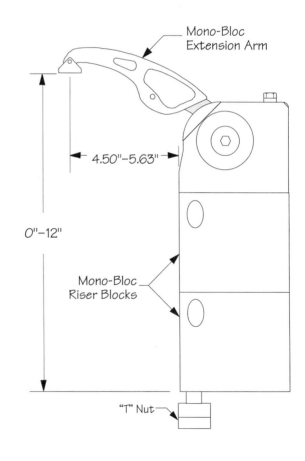

# Techniques for Clamping

**Description of Problem/Requirement:**
Design a fast-acting clamping system.

**Suggested Solution:**
Specify a cam clamp design for the workholder.

**Source:**
Various jig and fixture component manufacturers and in-house fabrication.

**Old Method:**

Eccentric cams are the simplest type of cam design and the most common for in-house fabrication. They apply the clamping pressure with the action of eccentric circles. These cams have a mounting hole positioned off-center in the cam lobe. The off-center location of the mounting hole produces the rise in the cam. The radial movement of the cam through its clamping cycle is the throw of the cam.

Special care must be exercised when using eccentric cams. Eccentric cams only have one true locking point. This is the point at which the vertical center lines of the eccentric mounting hole and the lobe are perfectly aligned and exactly perpendicular to the clamped surface. Any other cam position cannot provide a positive lock. Some form of mechanical lock, such as a hook or similar lock, should be used with any eccentric cam whenever vibration is expected.

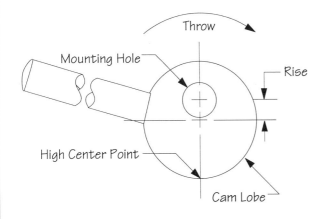

**New Method:**

Flat spiral cams, although similar to eccentric cams, differ substantially in their operational principles. These cams use an involute curve to provide a range of clamping positions instead of just one. The clamping range is produced by the curve of the cam. The flat spiral design also has a much smaller cam radius than an eccentric cam. An eccentric cam must have a radius 150% the size of a flat spiral for the same holding force. When clamping, the flat spiral cam requires only 63% of the force needed to lock an eccentric cam. These cams are commercially made as either single- or double-lobe cams.

The flat spiral design, although better than an eccentric cam, is still subject to loosening when vibrated. For this reason, a mechanical lock is also a good idea for applications where vibration may occur. Also, a double-lobe cam is usually more secure than a single-lobe type.

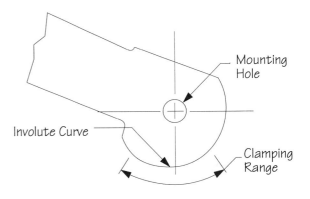

*Description of Problem/Requirement:*

Minimize the effects of vibration when using a cam clamp.

*Suggested Solution:*

Design the clamp assembly around indirect-pressure double-lobe cams.

*Source:*

Various jig and fixture component manufacturers and in-house fabrication.

*Old Method:*

Cams are typically used in two styles of clamping—direct pressure and indirect pressure. Direct pressure cams apply the force directly to the workpiece; indirect pressure cams apply the clamping force through a secondary clamping element. The main objections to direct-pressure clamps are the susceptibility of the clamps to vibration and the marring of the workpiece. The indirect pressure clamps are less prone to problems, but heavy vibration remains an issue.

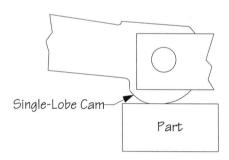

The cam applies the clamping pressure directly to the workpiece.

*New Method:*

The best combination of cam and application is a flat spiral cam design with an indirect pressure application. Likewise, the most suitable application for a cam clamp is where workpiece vibration is minimized. Drill jigs, light- to medium-duty milling fixtures, inspection fixtures, and assembly tools are all safe, effective applications for cam-type clamps.

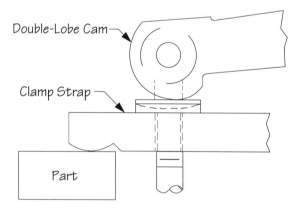

The cam applies the clamping pressure through another mechanical element.

# Techniques for Clamping

**Description of Problem/Requirement:**
Provide a fast-acting cam-type edge clamp that can be set up in multiple configurations.

**Suggested Solution:**
Incorporate a cam edge clamp into the workholder design.

**Source:**
Various modular fixture component manufacturers.

**Old Method:**

Standard edge clamps were formerly used to hold the workpieces for a particular fixture. However, since these standard edge clamps are operated with a screw thread, they were not fast enough to meet the scheduled production requirements. Rather than redesigning the complete workholder, the screw-type edge clamps were simply replaced with a cam-type edge clamp.

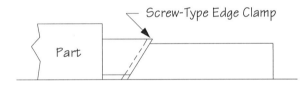

**New Method:**

The cam edge clamp is a clamp variation that applies the clamping action of the cam horizontally rather than vertically. The cam in this clamp is a flat spiral design having a 180° throw and a 5/32" range of movement. The clamp also applies the holding force indirectly, through the pivoting nose element. This makes the clamp more resistant to vibration. The pivoting nose element of the clamp applies the clamping force both forward and downward to securely hold the workpiece. These clamps may be mounted directly to the tool body, or on a variety of standard adapter bars.

**New Method, continued:**

## Cam Edge Clamp

## Mounting Cam Edge Clamps

*Description of Problem/Requirement:*

Design a fast clamping device that holds and locates from the surface to be machined.

*Suggested Solution:*

Specify up-thrust clamps for the workholder.

*Source:*

Various modular fixture component manufacturers.

*Old Method:*

The design of some odd-shaped workpieces can prevent them from being located and clamped in a conventional fashion. The workpiece shown here has a bottom surface that cannot be used to locate the workpiece for operations performed on the top surface. So, here, the workpiece must be both located and machined on the top surface. One type of clamp that can simplify the setup of this type of workpiece is the up-thrust clamp.

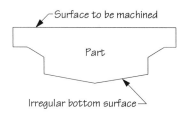

*New Method:*

The up-thrust clamp is a unique clamp design. Rather than clamping down, this clamp holds a workpiece with the clamping pressure applied from below, pushing up. Pushing the cam handle down moves the clamping element upward against the workpiece. The rotatable jaw element has two clamp openings—one for thinner workpieces and another for thicker workpieces. Rotating the cam screw moves the lower jaw up or down to precisely adjust the clamping thickness. The underside of the top jaw element has a precision ground surface and acts as a precise locator for the clamped workpiece.

*New Method, continued:*

## Up-Thrust Clamp

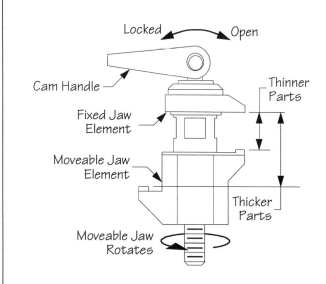

## Up-Thrust Clamp Setup

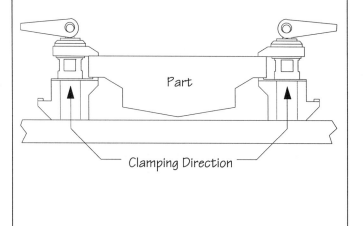

# Techniques for Clamping

**Description of Problem/Requirement:**
Design a fast-acting edge clamping device to securely hold a workpiece.

**Suggested Solution:**
Incorporate a wedge action clamping device into the workholder design.

**Source:**
In-house fabrication.

**Old Method:**

Edge clamping operations may be accomplished in many ways. One of the most common methods employs a screw thread to apply the clamping force. While effective, a screw clamp is not the most efficient for these operations.

**New Method:**

Many times, a wedge action clamp is a better choice for these clamping operations. Wedge clamps are some of the oldest clamping devices used in industry. Although not as widely used as other clamp forms, the wedge action principle is still adapted in various other workholding devices.

Most wedges for clamping are either flat wedges or conical wedges. Wedges may also be classified as either self-holding or self-releasing. Self-holding wedges are those that have a clamping angle of less than 4°. Self-releasing wedges have a clamping angle greater than 10°.

Flat Wedge

Conical Wedge

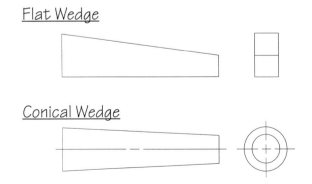

**New Method, continued:**

Self-Holding Wedge

Less than 4°

Self-Releasing Wedge

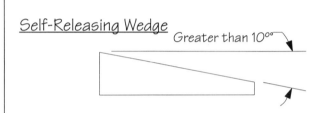

Greater than 10°

One clamping arrangement uses two wedges: one stationary and one moveable. The stationary wedge and the fixed jaw element provide the fixed reference points for the clamping. The moveable wedge, acting between the stationary wedge and the workpiece, applies the clamping force. The wedge is self-releasing (over 10°), so a threaded fastener is used as a mechanical lock.

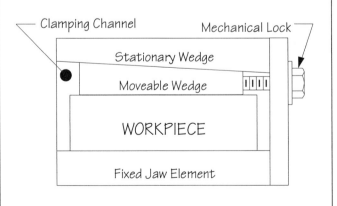

*Description of Problem/Requirement:*

Design a clamping device to hold and locate a workpiece in a hole.

*Suggested Solution:*

Incorporate a conical wedge into the workholder design as a clamping device.

*Source:*

Various jig and fixture component manufacturers and in-house fabrication.

*Old Method:*

Clamping inside a bored hole is an operation frequently required for some workpieces. One way this can be performed is with a mandrel turned to the precise diameter of the hole. This turned diameter locates the workpiece while a thread on the end of the mandrel combined with a nut, or screw, is used as a clamping device. Although very efficient, even slight differences in the hole diameter can greatly affect the overall locating and clamping accuracy.

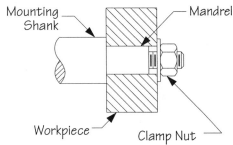

*New Method:*

Conical wedges are another form of wedge action clamp frequently used for holding some workpieces. These conical wedge clamps, also called tapers, like flat wedge clamps are either self-holding or self-releasing, depending on the angle of taper. Tapers less than 4° are self-holding, and those greater than 10° are self-releasing. The Morse taper shank found on cutting tools is an example of a self-holding conical wedge. The conical portions of standard 5C and R8 collets, and the taper in a milling machine spindle, are examples of self-releasing conical wedges.

*New Method, continued:*

All conical wedge clamps are self-centering. This characteristic makes them well suited for many workholding applications. Likewise, since the basic shape of these mandrels is conical, they can compensate for minor differences in the hole diameters from one workpiece to the next. The two general types of mandrels used for workholding are the solid mandrel and the expanding mandrel.

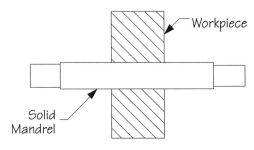

Solid mandrels have a taper angle of approximately 0.006" per foot.

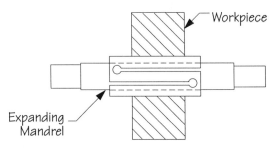

Expanding mandrels have various sizes and taper angles, depending on the manufacturer. Most will expand between 0.06" and 0.13" depending on the size.

# Techniques for Clamping 111

*Description of Problem/Requirement:*
   Design a universal clamping system to suit a wide range of clamping applications.

*Suggested Solution:*
   Specify toggle clamps in the workholder design.

*Source:*
   Various jig and fixture component manufacturers.

*Old Method:*

Any number of different style clamps are available for holding workpieces in a jig or fixture. However, toggle clamps are the most widely used clamping device. Few clamps offer the overall versatility and efficiency of the toggle clamp. Although they may not be appropriate for every application, in general, toggle clamps offer a wide range of benefits and should be considered for all suitable clamping situations.

The major benefits of the basic toggle clamp are an exceptional ratio of holding force to application force, rapid operation, positive locking action, and the ability to be applied in small or confined areas.

Toggle clamps operate on a system of fixed pivots and levers. The clamping action is performed through a series of fixed length levers connected by pivot pins. When in the unclamped or released position, the outer pivot points are retracted. This action retracts or raises the clamp off the workpiece when released. As the clamp is activated, the pivot points are extended, and pressure is exerted at both ends of the linkage. In the locked position, the linkage moves to its center position and snaps over the center point to lock the linkage. This sets the levers slightly beyond the center point and affords the positive locking action of the clamp. The clamping pressure lost when the linkage is moved beyond center is negligible.

*New Method:*

<u>Toggle Clamp Operation</u>

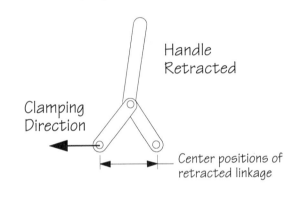

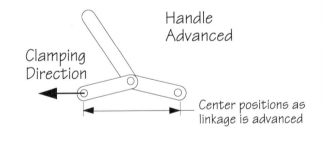

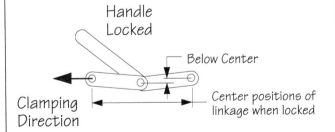

*Description of Problem/Requirement:*
Design a universal clamping system to suit a wide range of clamping applications.

*Suggested Solution:*
Specify toggle clamps in the workholder design.

*Source:*
Various jig and fixture component manufacturers.

*Old Method:*

The main reason toggle clamps are so versatile and adaptable is the wide range of commercially available toggle clamp designs. The four general clamping actions for standard toggle clamps are hold-down action, push/pull action, latch-action, and squeeze action.

<u>Hold-Down Action Clamps.</u> The hold-down action is the most common style of clamp. These are made with either a vertical or horizontal handle. They apply a downward clamping force.

<u>Push/Pull Clamps.</u> The push/pull clamps have a straight-line toggle action. These clamps are for either push- or pull-type clamping operations. The clamp applies clamping force at either end of the clamp stroke.

<u>Latch-Action Clamps.</u> The latch-action clamps operate on a pull action. These are often used with a latch plate to pull elements together.

<u>Squeeze Action.</u> The squeeze action clamps, also called toggle pliers, hold parts together with a squeezing action. These clamps work much like a toolmaker's clamp or a C-clamp.

*New Method:*

<u>Toggle Action Clamps</u>

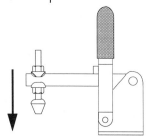
<u>Hold-Down Action</u>

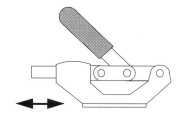
<u>Push/Pull Action</u>

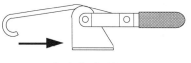
<u>Latch-Action</u>

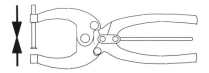
<u>Squeeze Action</u>

# Techniques for Clamping

**Description of Problem/Requirement:**
Design a toggle clamping system to suit a wider range of clamping thicknesses.

**Suggested Solution:**
Specify automatic toggle clamps in the workholder design.

**Source:**
Carr Lane Manufacturing Company.

**Old Method:**

For all their many benefits, the basic toggle action has a limited range of movement for different workpiece heights. Once set to a height, the standard toggle action can accommodate only minor thickness variations. Larger variations require readjusting the clamp spindle. The limitation is inherent in the basic design of the toggle action.

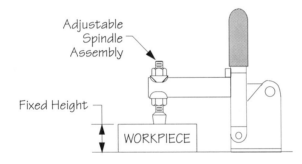

**New Method:**

The automatic toggle clamp has an internal self-adjusting feature to automatically adjust to different workpiece sizes. The clamp can accommodate variations in clamping heights of up to 15°. It has a total automatic-adjusting range of over 1.25", with constant and consistent clamping pressure. Additional adjustment, when needed, can be made with the clamp spindle. In use, the clamp is first set to the average workpiece height. The clamp then automatically adjusts for both smaller and larger workpieces.

**New Method, continued:**

## Automatic Toggle Clamp

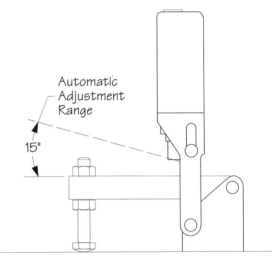

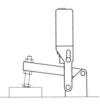
Set contact spindle to the average workpiece height

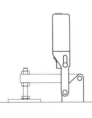
Clamp automatically adjusts to suit thinner workpieces

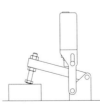
Clamp automatically adjusts to suit thicker workpieces

*Description of Problem/Requirement:*

Design a clamping system for holding very thin workpieces in multipart setups.

*Suggested Solution:*

Incorporate a subplate/epoxy system for mounting the very thin workpieces.

*Source:*

Various adhesive manufacturers or chemical suppliers.

*Old Method:*

Holding very thin workpieces is often a tricky task. However, holding several very thin workpieces in a multiple part setup is even more difficult. One common method of performing this type of operation is with a machine vise. Although adequate for some setups, when more than two workpieces are needed for each setup, or when the workpieces are very thin, this method may be less than satisfactory.

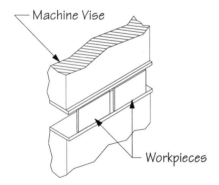

*New Method:*

An alternate method that may work for some situations is a subplate/epoxy clamping arrangement. Here, the workpieces are attached to the subplate with an epoxy, or other adhesive, rather than mechanically. This arrangement allows five sides of each workpiece to be machined in a single setup.

*New Method, continued:*

### Clamping Thin Workpieces with Epoxy

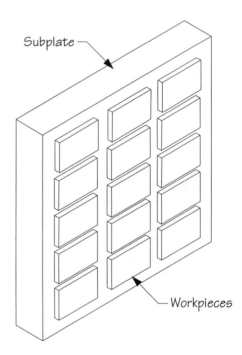

With adhesive clamping, the workpieces are mounted to the subplate with epoxy or another type of adhesive that can be removed with a solvent. The solvent must be on hand before attaching the workpieces to the subplate. This setup is well suited for any number of workpieces, and allows five sides of each workpiece to be machined. Only the subplate size will limit the number of workpieces per setup.

# Techniques for Clamping

**Description of Problem/Requirement:**
Design a clamping system to mechanically hold small or intricate workpieces.

**Suggested Solution:**
Incorporate the Grip Strip Tooling System into the workholder design.

**Source:**
Morgan Enterprises.

**Old Method:**

Edge clamping small or intricate workpieces can be very difficult. Most commercial edge clamps lack the correct form—or are too large—to suit the workpiece or the setup.

One way around this problem is the Grip Strip Tooling System. These units combine the design flexibility of custom-made clamps with the utility and low cost of standard clamps. The basic unit is a laminated bar consisting of a steel mounting plate and aluminum clamping plate bonded to a resilient intermediate layer of rubber. Rails on both sides of the clamping plate both clamp the workpiece and mount the unit.

In use, the unit is cut to length and a recess is machined in the aluminum plate to suit the shape of the workpiece. Here, the same computer program used to machine the workpiece can often be used for this recess. Simply set the cutter path to cut the inside periphery rather than the outside.

Single or multiple workpieces may be mounted in these fixtures depending on the length of the strip and the number of recesses. Clamp screws inserted in the rails are used to clamp the workpiece. Other devices, such as toggle clamps or power clamps may also be used. The units are also reusable—simply remachine the recess to a larger size.

**New Method:**

### Grip Strip Tooling System

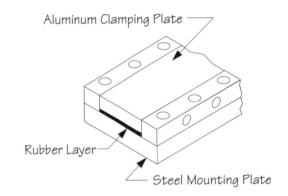

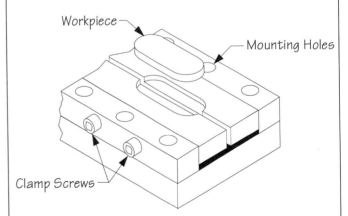

Bars are 18" long and in widths of 2" to 8". The aluminum clamping plates are made in three thicknesses: 0.25", 0.50", and 0.75".

**Description of Problem/Requirement:**

Design a clamping system to hold very large workpieces.

**Suggested Solution:**

Incorporate a claw clamp into the workholder design.

**Source:**

James Morton Inc.

**Old Method:**

Mounting and clamping large workpieces to angle plates, or similar fixturing elements, where there is no built-in clamping device can often present a host of problems. One of the more common methods of mounting these workpieces is with a conventional "C" clamp. Although adequate, these clamps are usually very large and hard to position. Likewise, the basic design of these clamps offers a limited range of sizes, so many different size clamps must be kept on hand to hold a variety of different size workpieces.

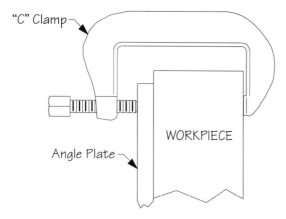

**New Method:**

One alternative to the conventional "C" clamp is the claw clamp. These clamps are much lighter and more versatile than the standard "C" clamp design. The claw clamp is available in several sizes with a range of up to 0"–40". The clamping jaw of these clamps is also reversible and may be flipped over to apply a reverse clamping action.

**New Method, continued:**

### Claw Clamp Setups

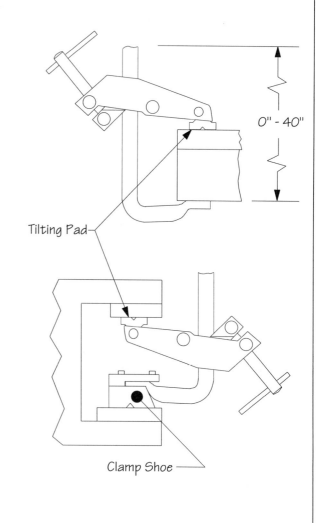

# CHAPTER SIX

# Techniques for Workholder Setups

## Techniques for Workholder Setups

**Description of Problem/Requirement:**
Provide an accurate and secure method of positioning the workholder on the machine table.

**Suggested Solution:**
Incorporate fixture keys into the workholder design.

**Source:**
Various jig and fixture component manufacturers.

**Old Method:**

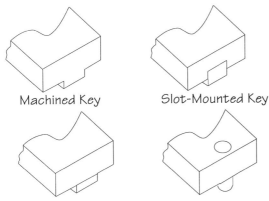

Machined Key   Slot-Mounted Key

Surface-Mounted Key   Pin-Type Key

Establishing the proper workholder location on the machine tool is vital to the accuracy of the operation. Simply bolting the workholder to the machine table is not recommended since the friction between the workholder and the table will not maintain the proper alignment. A better method of providing a positive reference when mounting workholders is with fixture keys.

Fixture keys may be made as part of the workholder base or designed as an additional element. When made as part of the base, the fixed size and loss of accuracy as the key wears are problems. Another problem is the additional cost of materials and labor. Even when the keys are installed as separate elements, the additional labor necessary to make the keys is a concern.

**New Method:**

A better alternative to the custom-made fixture keys are the commercially available keys. The commercially available fixture keys are much less expensive to buy than to custom fabricate. Commercial fixture keys also require much less effort to install. These keys are made in two general styles: slot-mounted and hole-mounted. Both styles are available in a variety of different sizes to suit the standard table "T" slots, and are normally case hardened to resist wear. These keys are typically mounted in either milled slots, or reamed and counterbored holes. Both key styles are attached with a single screw.

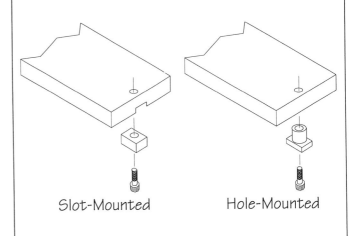

Slot-Mounted   Hole-Mounted

**Description of Problem/Requirement:**

Provide a method of locating workholders on tables with different size "T"-slots.

**Suggested Solution:**

Use step-type fixture keys to accommodate the different size "T"-slots.

**Source:**

Various jig and fixture component manufacturers.

**Old Method:**

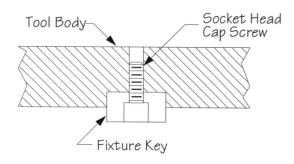

Both custom made and commercial fixture keys are very useful in providing a positive reference for locating workholders. However, despite their utility, both types are usually intended to locate workholders in a single size "T"-slot. Generally, the size of the fixture keys for a workholder is determined by the size of the "T"-slots in the machine table.

When another machine tool is required for the workholder, the table "T"-slot sizes must match the fixture key in the workholder. If these do not match, the key is usually removed to accommodate the substitute machine tool. This can affect the accuracy of the setup and the precision of the operation.

**New Method:**

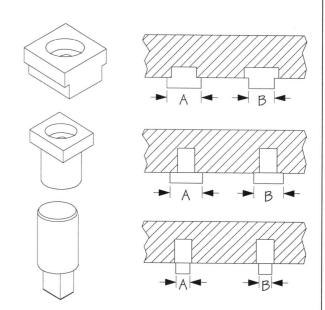

One way to adapt the fixture key to suit another machine tool table is with step-type fixture keys. Step-type fixture keys are a modification of the standard fixture key that allows a fixture to be used with a machine table having a slot width different than the workholder.

These step-type fixture keys are available in a variety of sizes in either slot-mounted or hole-mounted variations. The slot-mounted type may be used for table "T"-slots either larger or smaller than the fixture slot size. The hole-mounted type may be used to suit either of two different "T"-slot sizes by simply turning the key 90°.

# Techniques for Workholder Setups

**Description of Problem/Requirement:**

Develop a faster and easier method of mounting the fixture keys in the workholder.

**Suggested Solution:**

Use the expanding mount-type fixture keys.

**Source:**

Various jig and fixture component manufacturers.

**Old Method:**

Standard, commercially available fixture keys offer a variety of advantages over the custom fabricated fixture keys. However, despite their many advantages, even these keys are often time consuming to install. Depending on the style used, either a milled slot or a reamed and counterbored hole are required. Likewise, both variations require a drilled and tapped hole for the mounting screw.

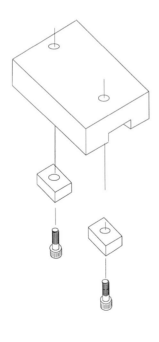

**New Method:**

The expanding mount-type fixture key is another variation that offers many cost-saving advantages. Rather than requiring multiple machining operations to prepare the mounting area, these fixture keys only require a drilled and reamed hole. To mount the key, the shank is simply inserted in the reamed hole, aligned to the table "T"-slot, and locked in place with a center mounted screw. This screw expands the shank, locking the key in the hole. This design allows the key to be secured from either the top or bottom of the workholder.

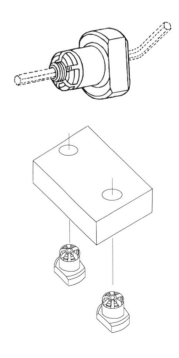

## Description of Problem/Requirement:

Quick method to load and locate a workholder on a machine table.

## Suggested Solution:

Use a modified "T"-nut and dowel pin arrangement.

## Source:

In-house fabrication.

### Old Method:

The former method incorporated two locating blocks, positioned at a right angle, and attached directly to the machine table. These blocks located the baseplate by one corner. This setup method required the edges of all the baseplates to be machined, adding to the overall cost of the workholders. During the machining operation, the baseplate had a tendency to slip away from the locating blocks, since only friction held the baseplate against the table surface. This resulted in locational inaccuracy and several spoiled workpieces.

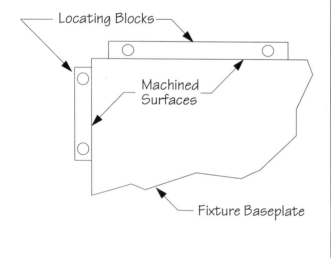

### New Method:

An alternative mounting method for these baseplates uses a modified "T"-nut arrangement. This "T"-nut has a dowel pin installed at one end and a set screw at the other. In use, the "T"-nut is held in place by tightening the set screw. The dowel pin extends above the table mounting surface and locates the workholder by engaging in two locating holes in the baseplate. To minimize the effects of wear, these locating holes should be fitted with drill bushings.

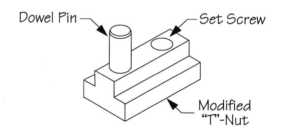

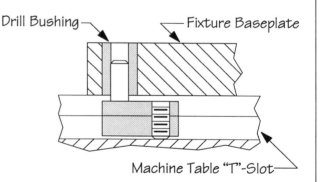

# Techniques for Workholder Setups

**Description of Problem/Requirement:**
Develop a method to load and locate workholders with different size mounting holes.

**Suggested Solution:**
Use a modified "T"-nut and dowel pin arrangement with interchangeable elements.

**Source:**
In-house fabrication.

**Old Method:**

Although the modified "T"-nut arrangement is quite useful for many location purposes, it does have one obvious design problem. Each of these units typically has a fixed size locator. This may not present a problem if all the baseplates have the same size locating holes. However, if several different size locating holes are used for different size baseplates, several of these modified "T"-nut devices might be required. The numbers of these units necessary may be further increased when a variety of table "T"-slot sizes are considered.

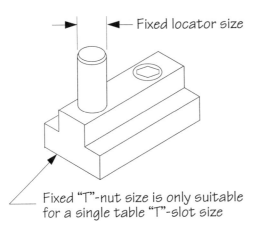

Fixed locator size

Fixed "T"-nut size is only suitable for a single table "T"-slot size

**New Method:**

A simple design modification can provide an alternate method of using this type of locator for a variety of different size baseplate mounting holes. Here, a variety of different size dowels and bushings may be used for each modified "T"-nut. With this arrangement, a range of different size locators may be used with a single "T"-nut. Although a variety of "T"-nuts are still needed to suit the various table "T"-slot sizes, this arrangement does substantially reduce the numbers of these units required for every possible workholder and machine table combination.

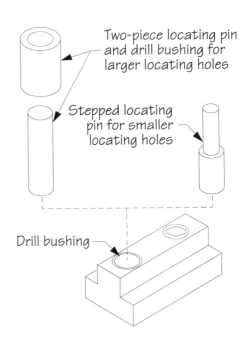

Two-piece locating pin and drill bushing for larger locating holes

Stepped locating pin for smaller locating holes

Drill bushing

| |
|---|
| *Description of Problem/Requirement:* |
| Design an accurate method to load and locate workpieces directly on a machine table. |
| *Suggested Solution:* |
| Use a modified "T"-nut and dowel pin arrangement. |
| *Source:* |
| In-house fabrication. |

*Old Method:*

Workpieces may be mounted in any number of different ways. Quite often, the size of the workpiece will determine how it is mounted. Larger workpieces are typically mounted directly on the machine table. Although a set of custom made locating rails, or blocks, may be used to position and locate these workpieces, they are not efficient for every type of workpiece.

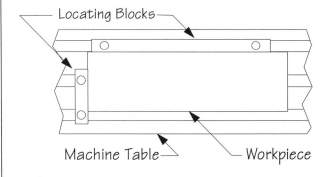

Quite often, these locating rails, due to their fixed length, are either too long or too short to suit the workpiece. Also, since the contact area between the workpiece and the locating blocks is often very large, chips and debris may cause locational problems.

*New Method:*

An alternative locating method for these larger workpieces is with the doweled "T"-nuts. Here, rather than locating the workholder, these units are positioned to locate the workpiece. In most cases, only three of these units will be necessary to locate any workpiece.

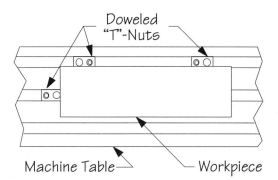

These locating units are positioned to restrict the horizontal degrees of freedom. Since each of these doweled "T"-nuts is positioned as a separate unit, the problem of dirt, chips, or other debris is minimized.

# Techniques for Workholder Setups

**Description of Problem/Requirement:**

Design a simplified locating unit to position small workpieces or workholders on a machine table.

**Suggested Solution:**

Use a right angle locator arrangement.

**Source:**

In-house fabrication.

**Old Method:**

Small workpieces or workholders often present a challenge when they are loaded directly to a machine table. Although a set of smaller size rails, or locating blocks, may be used, many of the problems inherent with the larger types are also present with these. Likewise, the time required to accurately position these elements is often excessive.

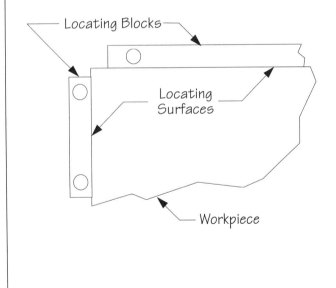

**New Method:**

A simplified locating device for mounting these workpieces is the right angle locator. This unit may be positioned almost anywhere on a machine table. This design provides a preset location in both horizontal directions. Depending on how this unit is made, either (or both) the inside and outside right angled surfaces may be used for locating a workpiece, workholder, vise, or other fixturing device.

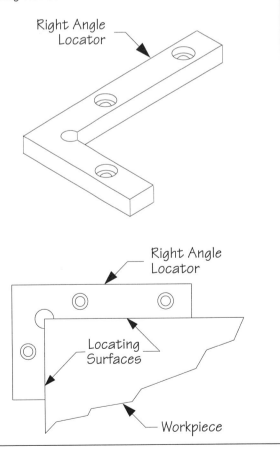

*Description of Problem/Requirement:*

Develop a universal positioning method to locate workpieces or workholders on machine tables.

*Suggested Solution:*

Use a single-axis locator unit.

*Source:*

In-house fabrication.

*Old Method:*

In most cases, when mounting workholders or workpieces directly on a machine table, the table "T"-slots are often used as a locating guide. There are, however, situations where either the shape of the mounted unit or the position of the table "T"-slots make this very difficult. Here a variety of methods may be used to provide a fixed reference point. One of the more common is to simply mount a clamp strap directly to the machine table in the necessary position.

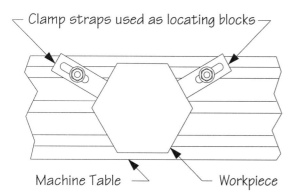

This method is very useful for lightweight setups or those with light machining forces. However, since these clamps are only held in place with friction, they may not be as suitable for heavier setups or those with greater machining forces. These conditions might cause the clamp straps to move.

*New Method:*

For heavy-duty setups, the single-axis locator unit might be a better choice. This design offers a larger contact surface and is capable of withstanding greater forces. Since these units are also fitted with fixture keys, they also have a positive reference in the table "T"-slot that reduces the tendency to move. The expanding shank-style fixture keys should be used here to accommodate the various positions of the locator.

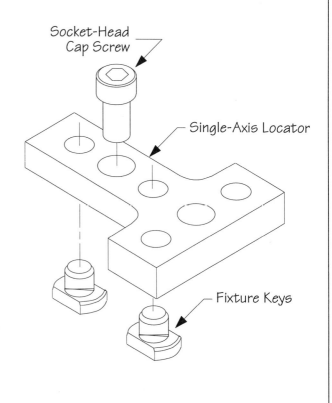

# Techniques for Workholder Setups

**Description of Problem/Requirement:**
Design a wear-resistant support to elevate the workholder off the machine table.

**Suggested Solution:**
Use a commercially made workholder rest.

**Source:**
Various jig and fixture component manufacturers.

**Old Method:**

Mounting supports are used to provide a stable mounting surface for the workholder. Many times, workholders are made without any support. This can cause problems if chips are under the workholder. The supports may also be machined directly into the base element. While better than no support, machined supports add significantly to the overall cost of the workholder; and when they wear, the entire tool body may need to be replaced. The best method is to use an installed support. Hex head screws are often used for this purpose. These screws, however, lack the precision and hardness usually required for most workholders.

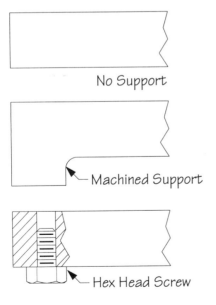

**New Method:**

Selecting the proper mounting supports is very important to the overall efficiency of any workholder. Whenever practical, commercial supports should be planned into the workholder design. These commercial supports are generally less expensive to buy and easier to install than custom supports. These supports are also hardened and ground, making them much more wear resistant and considerably more accurate than custom supports. Several types, styles, and sizes of supports are commercially available. The specific size and style of the supports selected is based on both the basic design and the intended functions of the workholder.

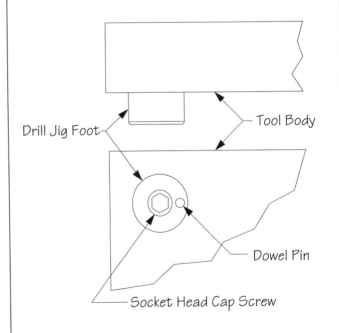

**Description of Problem/Requirement:**
Design an accurate and wear-resistant leveling device for mounting workholders on machine tables.

**Suggested Solution:**
Use an eccentric leveling lug.

**Source:**
Various jig and fixture component manufacturers.

**Old Method:**

Establishing and maintaining the correct relationship between the workholder and the machine tool is an important setup consideration. This frequently requires the workholder to be leveled. Several different methods and devices may be used to level workholders on the machine table. One common device used for leveling is a hex head screw mounted in a tapped hole. Once the workholder is properly leveled, a jam nut is used to lock the position. Although a very common method, this arrangement also presents a few problems. If the workholder is handled roughly, the screw can be damaged. Likewise, since the screws are seldom hardened, wear can be a factor that must be considered.

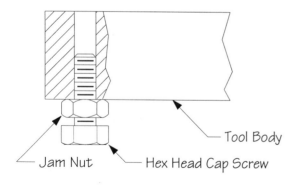

**New Method:**

One style of leveling device that is very useful for large and heavy workholders is the eccentric leveling lug. These levelers are less likely to be damaged if treated roughly, and since they are hardened, they are also wear resistant. Eccentric leveling lugs are mounted to the tool body through the counterbored mounting hole. The lug is then rotated to the desired height, and the mounting screw is tightened. The lug is then doweled in place to lock the unit in the desired location.

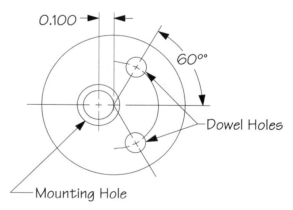

# Techniques for Workholder Setups

*Description of Problem/Requirement:*

Design an economic and wear-resistant device for raising a workholder off the machine table.

*Suggested Solution:*

Incorporate jig feet into the workholder design.

*Source:*

Various jig and fixture component manufacturers.

*Old Method:*

In designing some workholders, the area of the tool body where the workpiece is mounted must sometimes be elevated. Many times, if the height is not great, simply adding a machined pad might be all that is required. However, for those workholders where a substantial elevation is needed, adding additional material to the tool body might not be the most economical procedure. Several problems can arise when elevating members are machined into the tool body. The first is the additional cost of the material and labor needed to make these members. Similarly, if the workholder is used for a long time, these members might wear out. Here the workholder might need to be totally rebuilt, just to replace these legs.

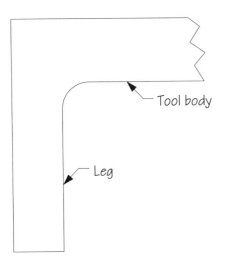

*New Method:*

A better way to add riser elements to a workholder is with jig feet. These commercially available units are made in two styles: single-end and double-end. Adding these to a tool body takes less material and considerably less labor. Likewise, although these units are hardened to resist wear, if they are damaged or wear out, they can easily be replaced, rather than rebuilding the complete workholder. Both styles of jig feet are available in a range of sizes and lengths. The single-end style is attached to the workholder with socket head cap screws, while the double-end type is mounted with a stud that connects both ends.

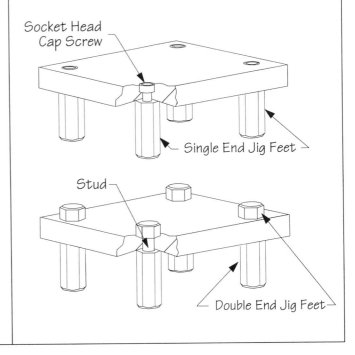

## Description of Problem/Requirement:

Design an economic alternative to threaded fasteners to reduce the cost of workholders.

## Suggested Solution:

Replace threaded fasteners with other devices in the workholder design.

## Source:

Various jig and fixture component and hardware manufacturers.

### Old Method:

Saving money is a primary goal in the design and construction of any workholder. As such, a designer should do anything possible to reduce the cost of the workholder. One way to do this is to reduce the labor expense. In the area of mechanical fasteners, the best way to cut expenses is to review the selection and installation of fasteners used for assembling workholders. Although mechanical fasteners do not cost very much, their installation is usually very expensive. Substantial savings can be achieved by rethinking the way fasteners are applied.

Screws, bolts, and nuts are common fasteners used for workholders. To mount these fasteners, the mounting area must be turned or bored to a specific diameter. Then, depending on the type of thread, either a tap or die is used to cut the required threads. Although these are simple operations and do not take much time, they do add an unnecessary amount to the labor cost of the workholder.

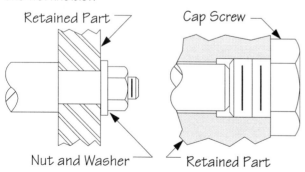

### New Method:

Other methods that can be used to perform many of the same functions in a workholder include using cotter pins or retaining rings. Cotter pins and retaining rings are inexpensive alternatives to using threaded fasteners. With a cotter pin assembly, the same functions are achieved at a lower cost. Here the parts only need to be drilled to accommodate the cotter pin. Retaining rings are just as simple to install. These alternatives may not work for every situation; however, where possible, they should be used to reduce workholder expenses.

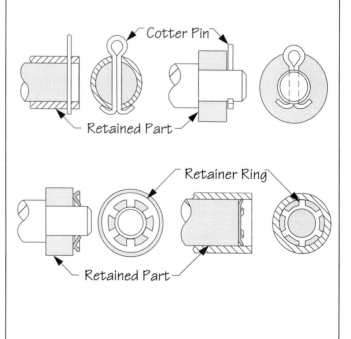

# Techniques for Workholder Setups

**Description of Problem/Requirement:**
Develop an inexpensive alternative to dowel pins to reduce the cost of workholders.

**Suggested Solution:**
Replace the dowel pins with spring pins or grooved dowel pins in the workholder design.

**Source:**
Various jig and fixture component and hardware manufacturers.

**Old Method:**

Dowel pins, like cap screws, are a common form of fastener used for assembling most workholders. The plain dowel pin is a hardened and ground pin with a radiused edge at one end and a slightly tapered lead angle at the other. In use, the mounting hole used for a dowel pin must be drilled and reamed to a precise size to allow sufficient interference between the dowel and the hole. Dowels are then pressed into the mounting hole and are held in place by the friction generated by the interference fit.

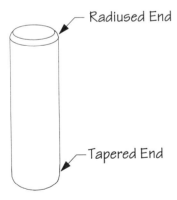

**New Method:**

In those cases where a workholder assembly does not require extreme precision, a less expensive alternative to the dowel pin may be used. Here an alternative type of locating pin might be a better choice. These pins are the spring pin (also called roll pin) or a grooved dowel pin. Both alternatives are less expensive to install because they only require a drilled hole. With these, the reaming operation is eliminated.

Spring, or roll, pins are made from flat stock and rolled into a cylindrical shape and hardened. In use, the pin compresses as it is installed, and the spring pressure of the pin holds it in the hole. Groove pins are solid dowels made with a series of grooves around their outer surface. When these are installed, the grooves bite into the hole and hold the pin in place. Neither of these is as accurate as a dowel pin, so they should only be used where the locational tolerances permit.

Spring Pin or Roll Pin     Grooved Dowel Pin

*Description of Problem/Requirement:*

Design an effective alternative to standard pull dowels for use in blind holes.

*Suggested Solution:*

Replace standard pull dowels with Metaligner two-piece pull dowels.

*Source:*

S.B. Whistler & Sons.

*Old Method:*

Standard dowel pins are widely used for building workholders. Despite their wide use, plain dowel pins do have a serious limitation: they are almost impossible to remove from a blind hole. Generally, dowel holes are machined completely through the parts to permit the dowel to be driven out from the reverse side. With some applications, however, the mounting hole cannot be machined through the parts. Here, pull dowels are used. These dowels can be installed and removed from the same side.

One popular pull design has a tapped hole in the end of the dowel to allow a screw and washer to be used to remove the dowel. When using pull dowels, remember that the air in the blind hole is compressed as the dowel is installed. To prevent problems in installing any dowel in a blind hole, always grind a small flat spot on the side of the dowel to allow the air to escape.

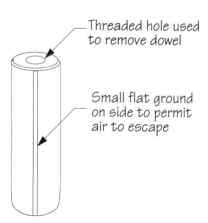

- Threaded hole used to remove dowel
- Small flat ground on side to permit air to escape

*New Method:*

When plain pull dowels are used, the mounting holes can wear out after repeated installation and removal of the dowels. To minimize this problem, a two-piece dowel assembly can be used. These dowels are not pressed into the hole, but rather the tapered sleeve is dropped in the hole and expanded with a tapered dowel. This action provides the necessary alignment without wearing out the mounting hole. These dowels are removed with a special pull nut and sleeve puller. The pull nut threads onto the end of the tapered dowel and breaks the tapered lock between the dowel and sleeve. The sleeve is removed from the hole with a special hook called a sleeve puller.

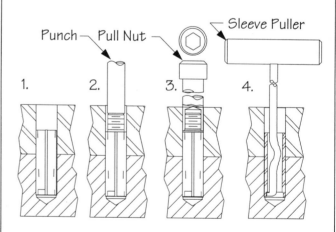

1. The sleeve is inserted in the reamed hole.
2. The tapered dowel is set with a punch.
3. The dowel is removed with a pull nut.
4. The sleeve is removed with a sleeve puller.

# Techniques for Workholder Setups

**Description of Problem/Requirement:**
Design an aligning device to position workholder elements.

**Suggested Solution:**
Incorporate alignment pins into the workholder design.

**Source:**
Various jig and fixture component manufacturers.

## Old Method:

Alignment pins serve a variety of functions with jigs and fixtures. Most commonly, these devices align the workpiece to the workholder or position removeable workholder elements. Alignment pins come in many types, styles, and sizes for many applications.

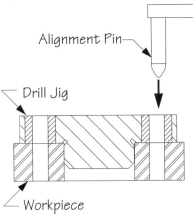

The simplest form of alignment pin is the plain alignment pin. These are available in a variety of sizes and styles. They are also made with several different handle designs. The most common handle designs are the "T" pin and "L" pin. The sliding handle pin is another design especially useful for areas where space is limited. Each of these pins has a bullet nose design.

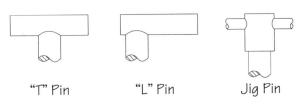

"T" Pin     "L" Pin     Jig Pin

## New Method:

The specific mount for the alignment pins is determined by the application. Pins are usually installed in drilled and reamed holes. For more precision or for longer production runs, they may also be installed in hardened bushings. Another style of bushing used when the workpiece alignment holes are slightly misaligned are the slotted locator bushings.

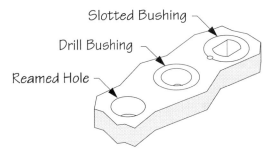

Alignment pins are also available in several end styles. The plain bullet nose pin is one of the more common; however, detent pins and clamping pins are also available. The detent pin uses spring-loaded balls to hold the pin in the hole. The clamping pin has a moveable bushing that is welded in place once the proper clamping distance for the pin is set.

Bullet Nose Pin     Detent Pin     Clamping Pin

**Description of Problem/Requirement:**

Design a device to keep the alignment pins with the workholder.

**Suggested Solution:**

Incorporate a locking pin into the workholder design.

**Source:**

Carr Lane Manufacturing Company.

**Old Method:**

Alignment pins are frequently used for a variety of different workholders. One problem that often occurs with these elements is simply keeping track of the pins when they are not engaged in a workpiece. If this problem is overlooked in the design, these pins can become lost or damaged. One way to make sure the alignment pins stay where they are needed is with a cable assembly. A cable assembly attaches the pin to the body of the workholder where it is easily accessible to the operator. When mounting a cable, make sure the cable is long enough to allow easy installation, but not so long as to interfere with the workholder or the machine tool.

**New Method:**

Although cable assemblies are quite useful for some workholders, another alternative is to include a locking pin into the workholder design. The locking pins are mounted in a special bushing. These bushings are attached to the workholder and securely attach the pin to the workholder. A spring clip ring in the locking pin bushing holds the pin in its retracted position. It engages in a groove in the end of the pin. The ring also puts spring pressure on the pin at any point along its length to hold the pin position when the workholder is moved or turned upside down. This locking arrangement is available with either alignment pins or clamping pins.

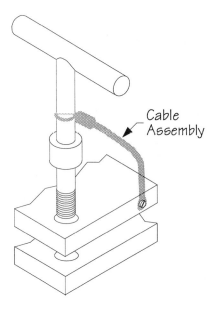

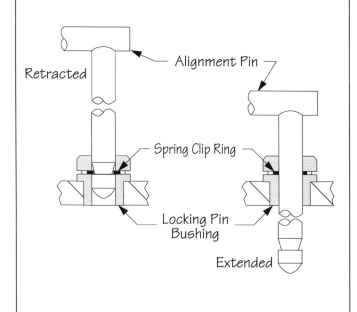

# Techniques for Workholder Setups

**Description of Problem/Requirement:**
Design a fast-acting device to clamp workholder elements together.

**Suggested Solution:**
Incorporate a quick-release-style alignment pin into the workholder design.

**Source:**
Carr Lane Manufacturing Company.

**Old Method:**

Many times, the workholder elements positioned with alignment pins must be held together as well as aligned. With these applications, clamping pins are often used to both hold and align the elements. To mount these pins, one of the workholder elements must be tapped to accommodate the threaded end of the clamping pin. While this arrangement may work very well for a variety of applications, there are situations where a tapped hole in one of the elements might be objectionable. Likewise, since the clamping action is accomplished with a threaded element, these clamping pins take more time to operate.

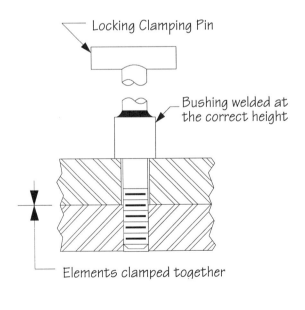

**New Method:**

The adjustable ball lock pin and the expanding pin also hold the elements together. However, since neither pin is threaded, both are faster acting than the clamping pin.

### Adjustable Ball Lock Pin

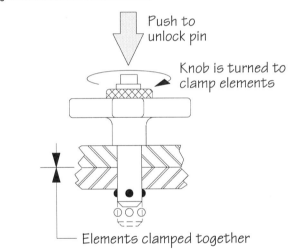

### Expanding Pin

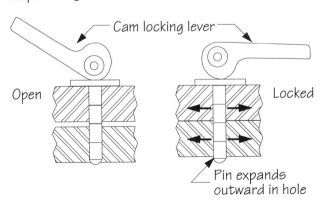

*Description of Problem/Requirement:*
Design a method to prevent alignment pins from wearing out their mounting holes.

*Suggested Solution:*
Incorporate hardened bushings into the workholder design.

*Source:*
Various jig and fixture component manufacturers.

*Old Method:*

Many of the pins used with workholders are designed to be inserted and removed regularly. These pins are generally used for aligning workholder elements or for the workpiece. However, the constant insertion and removal of these pins can often wear the mounting holes. Although hardened steel workholder elements may resist this wear, workholders made from soft steel, aluminum, or other softer materials cannot. To reduce the problem of wear and to maintain the accuracy of the workholder, hardened bushings should be used for these holes.

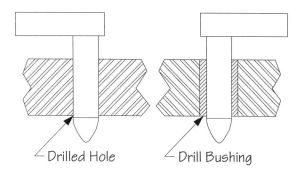

*New Method:*

Mounting bushings in steel workholder elements does not usually present a problem. However, softer materials such as aluminum or nonmetallic materials may not hold standard drill bushings as well as steel elements. Here, another style of drill bushing might be better suited. These are the knurled or serrated drill bushings.

Knurled drill bushings have a diamond pattern knurl on their outer diameter. This style bushing is best for cast-in-place applications. Here, the mounting hole is made slightly oversize, and the space between the bushing and the inside of the hole is filled either with an epoxy resin or a low-melt alloy. Serrated bushings have a straight pattern knurl on their outer diameter. These are intended for press fit applications.

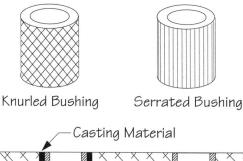

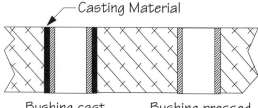

# Techniques for Workholder Setups

*Description of Problem/Requirement:*

Design a safe system for lifting and moving the workholders.

*Suggested Solution:*

Incorporate commercial hoist rings into the workholder design.

*Source:*

Various jig and fixture component manufacturers.

*Old Method:*

Any workholders weighing over 30 pounds should be fitted with hoist rings to allow them to be moved safely. When selecting hoist rings, one very popular type of device used for lifting purposes is a standard eye bolt. Although very common, eye bolts should only be installed on very lightweight workholders. Likewise, when lifting workholders with eye bolts, always make sure the lifting angle is as great as possible to reduce any side loads. Excessive side loads applied to a conventional eye bolt can cause the unit to fail.

*New Method:*

Commercial hoist rings are available in several types and styles. For safety reasons, only approved commercial hoist rings should be used for lifting any workholders. Always make sure the weight of the workholder is well within the rated capacity of the hoist ring. The specific style of hoist ring selected should be based on the work to be done. Most commercial-style hoist rings have a lifting ring that pivots 180°. Likewise, the swiveling-type hoist ring also rotates 360° to align itself to the lifting source.

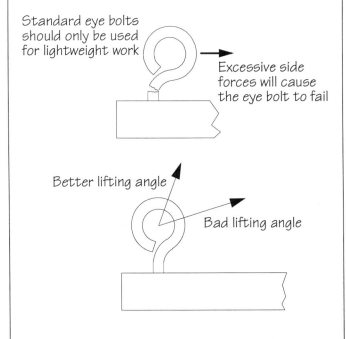

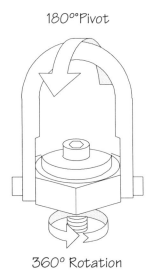

**Description of Problem/Requirement:**
   Develop an efficient and economical system for multiple machining operations on small workpieces.

**Suggested Solution:**
   Construct a simplified custom pallet system for small workpieces.

**Source:**
   In-house fabrication.

**Old Method:**

Low-volume production runs involving small workpieces are often fixtured in a variety of common shop workholding devices such as vises, chucks, or collets. Another alternative is a simplified custom pallet arrangement. With this arrangement, receiver units are made for each machine tool required for the operations. The workpieces are mounted on the carriers and moved between the various machine tools. Depending on the precision required, a simple dowel pin arrangement may provide the necessary repeatability between setups. Once mounted, the carrier is held on the receiver with a flange nut and washer.

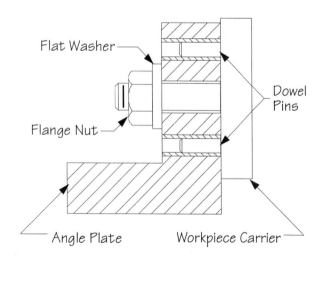

**New Method:**

Setups demanding more precision require another approach. A self-releasing tapered mount for the workpiece carrier and receiver unit offer greater precision. Here, the self-centering characteristics of the conical form assure perfect alignment of the elements each time they are assembled. A dowel pin can also be included to provide the radial alignment. The exact size of the tapered shank and hole, although not critical to the overall operation, is still important. Tapers with angles greater than 10° are self-releasing tapers. Those tapers with an angle of less than 3° are self-holding tapers. For this application, the taper angle should be greater than 10° to operate properly. A flange nut and "C" washer hold the carrier in place. This type of washer allows the complete carrier unit to be extracted without removing the nut.

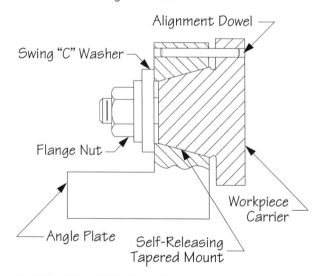

# Techniques for Workholder Setups

**Description of Problem/Requirement:**
Develop an efficient and economical alternative to dedicated workholders.

**Suggested Solution:**
Use major structural elements and the Master Plate method to simplify fixturing workpieces.

**Source:**
Various jig and fixture component manufacturers and in-house fabrication.

**Old Method:**

Many times, complete dedicated workholders are designed and constructed for each machining operation needed for a workpiece. Although this procedure may be required for some workpieces, frequently a less expensive alternative should be explored.

Where possible, workpieces should be fixtured with the wide range of commercially available vises, chucks, and similar workholding devices. Where the size, shape, or other parameters of the workpiece do not permit these devices to be employed, other commercial alternatives should be explored. Frequently, workpieces may be mounted directly to the variety of commercially available major structural elements. These include a wide array of subplates, angle plates, double angles, tombstones, and similar units.

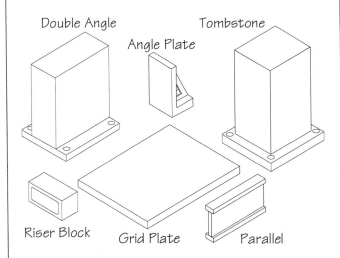

**New Method:**

The Master Plate tooling method uses a simple plate as an adapter between the workpiece and the major structural elements. Instead of mounting the workpiece directly to the fixturing elements, the workpiece is first attached to a master plate. This plate is then mounted to the structural elements. This method allows the workpiece to be fixtured for a variety of different operations and setups. Here the master plate acts as a pallet arrangement. The blank workpiece is first mounted to the master plate and cycled through all the operations. When the machining is completed, the finished workpiece is removed and replaced with a new blank workpiece.

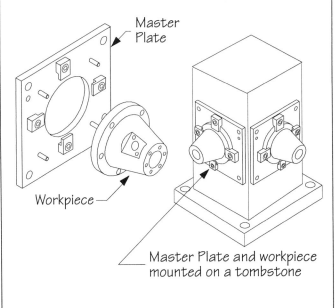

*Description of Problem/Requirement:*
  Develop an efficient and economical alternative to dedicated workholders.

*Suggested Solution:*
  Use individual structural elements or combinations of assembled elements to fixture workpieces.

*Source:*
  Various jig and fixture component manufacturers.

*Old Method:*

These major structural elements are made in two general styles: "T"-slot and dowel pin. The "T"-slot-type elements use a series of precisely machined "T"-slots to mount accessories. Generally, workpieces are easier to clamp with "T"-slot components. However, due to the limited number of fixed registration points, workpieces are more difficult to locate. The only fixed registration points are at the intersections of the "T"-slots.

Dowel pin components are similar in design to "T"-slot-type elements. The major difference is in the grid pattern of holes used to locate and mount other accessories. The major advantage to using dowel pin-type components is in the automatic positioning of the elements in the holes. Here, each hole is a fixed registration point. This large number of registration points makes locating simpler. However, since the holes are fixed, positioning the clamps is somewhat more cumbersome.

*New Method:*

When using either "T"-slot or dowel pin components, the major structural elements may be used alone, or combined to suit the requirements of the workpiece. Likewise, other workholding devices, such as vises, chucks, or similar elements, may also be mounted directly to these structural elements. It is always less expensive to use these components in place of custom dedicated fixturing elements made specifically for a workpiece. Not only are they less expensive to buy, but unlike dedicated components, these units are also reusable. Whenever practical, these components should be incorporated into the workholder design.

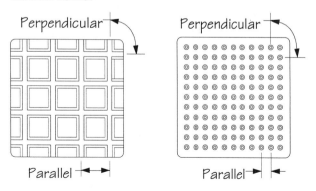

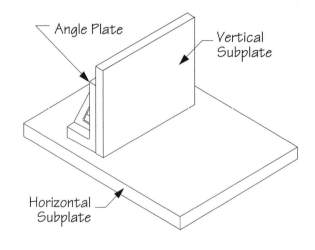

# Techniques for Workholder Setups

**Description of Problem/Requirement:**
Develop a fast and efficient method of mounting workholders.

**Suggested Solution:**
Incorporate the Ball Lock™ Mounting System into the workholder design.

**Source:**
Jergens Incorporated.

**Old Method:**

Workholders are attached to machine tables or subplates with a variety of fasteners. One of the more popular mounting methods is with screws and dowels. Here the subplate is often fitted with dowel pins and tapped holes. When mounted, the workholder is aligned with the dowels and then held in place with the screws.

Although this arrangement is very common, it is not without its problems. The dowels can be damaged through accident or carelessness during either the mounting operations, or while the machine is left idle. Likewise, the tapped holes must be kept covered or they can easily become packed with dirt, chips, or debris. Also, with repeated use, the threads will often wear out; this requires either a new subplate, or fitting the tapped holes with thread inserts.

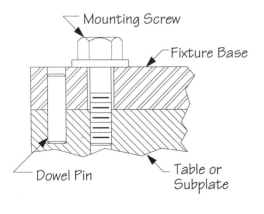

**New Method:**

An alternative to using a screw and dowel arrangement is the Ball Lock™ Mounting System. This system has three elements: a liner bushing, a receiver bushing, and a locating shank. The liner bushing is mounted in the fixture base. The receiver bushing is installed in the machine table or subplate. These bushings both centrally locate and lock the locating shank.

In use, the locating shank is inserted in the bushings and the locking screw is tightened. The screw contacts the large ball within the locating shank. As this ball is advanced, it engages three small locking balls that are moved outward in their sockets into the conical sides of the receiver bushing. The locking balls exert force against the conical portion of the bushing, locking the balls in position. This clamping action seats the shank by pulling it down into the bushing.

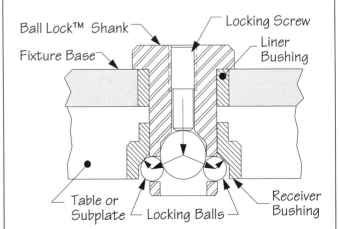

*Description of Problem/Requirement:*

Develop an efficient method of reducing workholder setup times.

*Suggested Solution:*

Standardize baseplate sizes and use the Ball Lock™ Mounting System to mount workholders.

*Source:*

Jergens Incorporated.

Old Method:

The simplest way to decrease the setup time from one job to the next is by standardizing the sizes and mounting methods of the workholders. When designing jigs and fixtures, for example, the normal practice is to make the baseplate, or tool body, large enough to accommodate the workpiece and the other fixturing elements. This results in a wide variety of different size jigs and fixtures. In fact, in most shops, virtually all the workholders have different overall sizes.

By establishing a limited series of standard baseplate sizes and hole patterns, the time needed to set up each workholder can be greatly reduced. The specific number of different size baseplates as well as the actual sizes of the plates is determined by the general sizes of the workpieces and fixtures used. But these standardized baseplates should not be limited only to new workholder designs—your existing workholders may also be retrofitted or mounted on standardized mounting plates.

New Method:

The Ball Lock™ Mounting System may be used to attach the fixture plate to the subplate. These units are faster and easier to use than a threaded fastener and dowel pin arrangement. They provide maximum mounting security and afford highly accurate and repeatable location. Ball Locks permit positional accuracy to within 0.0005" of true position. With their mounting arrangement, there are no projecting dowels, threaded fasteners, or tapped holes to pack with debris. The number of Ball Locks used for a workholder is determined by the size of the baseplate. Many times, only two Ball Locks are needed, but with larger workholders, four may be used. When four Ball Locks are used, only two will have liner bushings for alignment, the other two only hold the workholder.

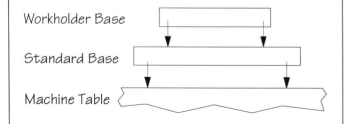

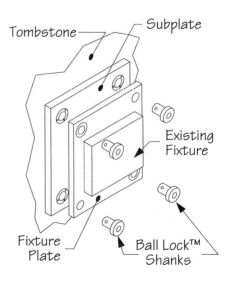

# Techniques for Workholder Setups

**Description of Problem/Requirement:**
Develop a quick-change fixturing system for existing workholders.

**Suggested Solution:**
Standardize baseplate sizes and use the Ball Lock™ Quick-Change Kit to mount workholders.

**Source:**
Jergens Incorporated.

**Old Method:**

Standardizing workholder setups can often result in substantially reduced setup times. Some of the more common methods of standardizing these setups involve using a collection of mounting devices such as subplates, double angles, and tombstones to mount existing workholders. These fixturing elements may then be combined to create a palletized system. Although these devices will definitely help reduce setup times, the methods used to mount the workholders should also be standardized to achieve the best setup times.

The dowel pin and threaded fastener methods of mounting workholders, although quite common, are time consuming when changing workholders. For this reason, when developing a quick-change fixturing system, other mounting alternatives should be explored to reduce the changeover times. Even a small amount of time saved in each individual step can result in a substantial collective savings over the complete workholder loading and unloading process.

**New Method:**

One quick-change fixturing system that offers a wide range of possibilities is the Ball Lock™ Quick-Change Kit. This system consists of a steel subplate with the receiver bushings preinstalled, aluminum fixture plates with the matching liner bushings preinstalled, and four Ball Lock™ shanks. The subplates are available in several sizes, in either single-plate or double-plate configurations. The fixture plates may be used as a base for the workpiece fixture, or as a secondary base for an existing fixture.

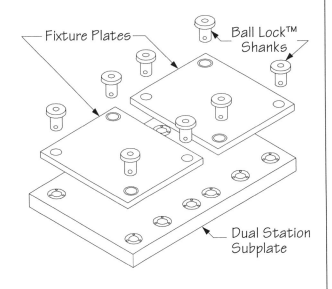

# CHAPTER SEVEN

## Techniques for Chucks and Collets

# Techniques for Chucks and Collets

*Description of Problem/Requirement:*

Develop a method of holding nonstandard workpieces in standard chucks.

*Suggested Solution:*

Replace the standard hardened chuck jaws with machinable soft jaws.

*Source:*

Various jig and fixture component manufacturers.

*Old Method:*

Chucks have long been used for a wide range of workholding applications. Although primarily used for lathe turning operations, chucks are also used for milling, grinding, and a host of other processes. The most commonly used styles of chucks are the 3-jaw and 4-jaw types.

The standard jaws furnished with these chucks are usually hardened and are well suited for holding a variety of standard workpiece shapes. The jaws on most standard chucks may set up to hold workpieces in three ways: normal, reversed, and internal.

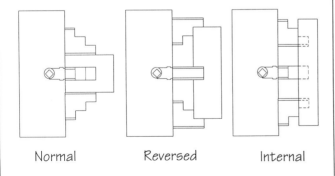

Normal        Reversed        Internal

*New Method:*

The standard hardened chuck jaws found on all standard chucks, although widely used for many workholding operations, are not always the best choice for all workpieces or all machining operations. For some workholding situations, a better choice is a set of jaws machined to suit the particular workpiece form or the operations being performed. Here a set of soft jaws is often a better choice. Soft jaws are typically made from aluminum, low carbon steel, or nonmetallic materials. These jaws may be machined to match the requirements of the workpiece or the machining operations. Soft jaws are made in sets to suit the specific type of chuck used for the operation. The mounting area where the jaw attaches to the chuck is premachined to suit the chuck. The gripping areas of the jaws are left unmachined and are usually machined in place to suit the workpiece.

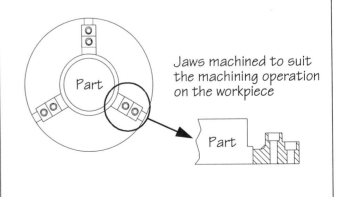

Jaws machined to suit the machining operation on the workpiece

## Techniques for Chucks and Collets

**Description of Problem/Requirement:**
Develop a method of holding thin-walled and irregular workpieces in standard chucks.

**Suggested Solution:**
Replace the standard hardened chuck jaws with machinable full grip soft jaws.

**Source:**
Various jig and fixture component manufacturers.

**Old Method:**

Some workpiece shapes are difficult, if not impossible, to hold in a standard chuck. Although machinable soft jaws will allow some of these workpieces to be fixtured, even these chuck jaws are not suitable for all workpieces.

Two general categories of workpieces that are not easily held in chucks, with either standard jaws or machined soft jaws, are thin-walled workpieces and odd- or irregular-shaped workpieces. Thin-walled workpieces have a tendency to deform when they are gripped with individual chuck jaws. As the chuck is tightened, the workpiece will spring out of shape. If bored in this condition, the hole will be round while the workpiece is clamped. Once released, however, the workpiece will return to its normal shape and the bored hole will be egg-shaped. Odd- or irregular-shaped workpieces are difficult to hold because of their shape. Although a 4-jaw independent chuck may be used for some shapes, the time required to align the workpiece is often excessive.

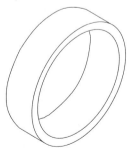
Thin-Walled rings

Noncylindrical Workpieces

**New Method:**

Full grip jaws are a form of soft jaw that make machining difficult workpieces much easier. These jaws are made in a variety of standard sizes in either aluminum or cast iron. Like standard soft jaws, they are machined to match the exact jaw mounts on the specific chucks.

Full grip jaws are made to cover the entire face of the chuck. These jaws are normally used for workpieces having odd shapes or those requiring more support. When used for thin-walled workpieces, a cavity is machined to suit the workpiece diameter. When the workpiece is mounted and clamped, the clamping force against the workpiece is distributed equally around the complete diameter, and the workpiece cannot deform or distort. With odd- or irregular-shaped workpieces, the cavity in the jaws may be machined on a milling machine rather than a lathe.

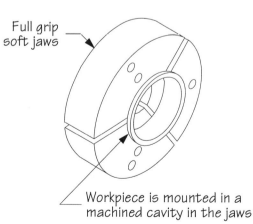
Full grip soft jaws
Workpiece is mounted in a machined cavity in the jaws

# Techniques for Chucks and Collets

*Description of Problem/Requirement:*

Design a method of holding cast or forged workpieces securely in standard chucks.

*Suggested Solution:*

Add carbide or hardened tool steel grippers to the clamping faces of the soft jaws.

*Source:*

Various jig and fixture component manufacturers.

*Old Method:*

Machinable soft jaws are very adaptable for a wide range of workpiece shapes and sizes. However, some workpieces are still very difficult to hold, even with custom jaws. Typically, cast or forged workpieces are difficult to hold securely due to their irregular mounting or clamping surfaces. This may not be a problem with slower speed operations where only light machining is performed. However, for operations requiring faster speeds and heavier cutting operations, the surface irregularities can affect the security of the setup.

One method of adding greater clamping security for these workpieces is by cutting an angular undercut to the jaws. This will help pull the workpiece back against the jaw and minimize any tendency of the workpiece to pull out of the chuck.

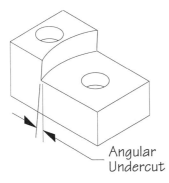

Angular Undercut

*New Method:*

Another method of adding security to the clamping operation, when cast or forged workpieces are held in a chuck, is with grippers installed in the gripping faces of the soft jaws. These grippers are made in several sizes in either hardened tool steel or carbide. These grippers are made with a diamond pattern gripping surface and are usually available with fine, medium, or coarse tooth patterns. When a workpiece is clamped with these devices, the teeth of the grippers penetrate the workpiece surface to afford additional clamping security. Since these grippers mar the workpiece surface, they should only be used for those applications where this marring is not objectionable or where the marks left will not affect the final product.

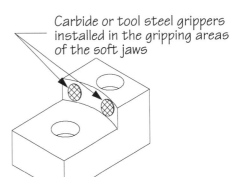

Carbide or tool steel grippers installed in the gripping areas of the soft jaws

*Description of Problem/Requirement:*

Develop a faster and more universal method for boring soft jaws.

*Suggested Solution:*

Use a universal forming device for boring the soft jaws.

*Source:*

Huron Machine Products.

*Old Method:*

When working with soft jaws, one operation that must always be performed before loading the workpiece is machining the jaws. One method of performing this machining is with a series of rings and plugs. The rings and plugs eliminate any backlash in the chuck, and provide the required resistance against the jaws to permit them to be machined accurately. Rings are used when the jaws are bored to a specific inside diameter. Plugs are intended for turning an outside diameter on the soft jaws. Since there is an infinite combination of workpiece diameters and different size chucks used in most shops, these rings and plugs are often machined for each set of soft jaws.

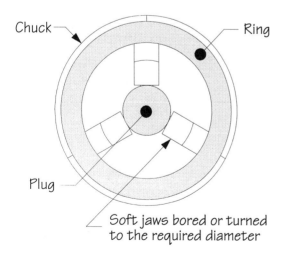

*New Method:*

An alternative to turning different rings and plugs for each job is the *Huron Forming Device*. This unit mounts in the counterbored mounting holes of the soft jaws and precisely holds the jaws in position, while eliminating the chuck backlash. This is a universal forming tool that can be used with a wide range of different chucks and turned or bored diameters.

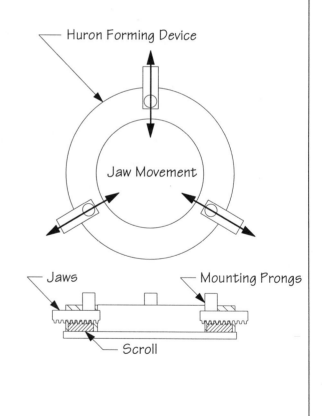

# Techniques for Chucks and Collets

*Description of Problem/Requirement:*

Develop a faster, less expensive, and more productive procedure for machining soft jaws.

*Suggested Solution:*

Use a universal turning fixture to machine the soft jaws.

*Source:*

Huron Machine Products.

*Old Method:*

One of the more common methods of machining soft jaws for any particular workpiece is to simply mount the blank soft jaws directly on the chuck that is intended to be used in performing the actual production machining operations. While this is a common practice in many shops, this method does tie up the production machines with nonproduction jobs. While machining these soft jaws, production machines cannot be used to make parts.

Another problem with this method of machining the soft jaws occurs when specialty jaw shapes are necessary. Here, the chuck may be removed from the production machine to permit the jaws to be machined in another machine tool. This can add a considerable amount of downtime to the operation, and further reduces the output of the production machine tools.

*New Method:*

A better method of turning soft jaws is with a turning fixture. These units simplify the procedures required to produce soft jaws. They are more economical to use than the production machine tools and allow the job of machining soft jaws to be done "off-line." With this fixture, the production machines can be used to produce parts instead of fixturing elements.

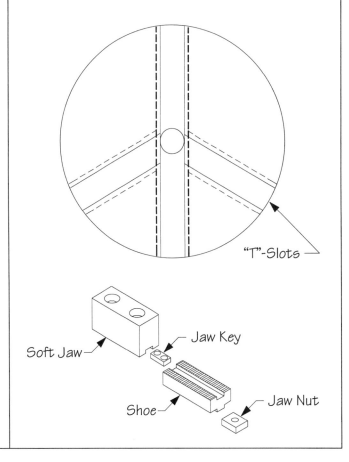

# Techniques for Chucks and Collets

**Description of Problem/Requirement:**

Develop a faster, less expensive, and more productive method of changing soft jaws.

**Suggested Solution:**

Use a quick-change mounting system for the soft jaws.

**Source:**

Huron Machine Products.

**Old Method:**

The primary advantage of using soft jaws for any machining operation is the ability to adapt the workholding element to the workpiece. Machining the jaws to suit the shape of the workpiece offers many production advantages. A primary advantage of employing this setup arrangement is that the jaws match the shape of the workpiece, so little or no adjustment is necessary when mounting the individual workpieces. This can save many hours of production time.

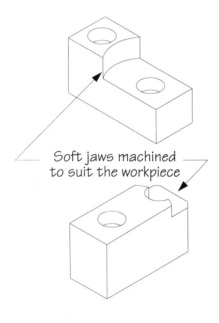

Soft jaws machined to suit the workpiece

**New Method:**

Another variation of the soft jaw idea that adds even more functionality to this inexpensive alternative to dedicated workholders is the quick change soft jaw system. The design of this quick change system uses a vertical dovetail to clamp the machinable soft jaw in place. These jaws are available in several heights and lengths and either hard or soft to suit a wide range of possible setups. The master jaw element stays mounted to the chuck while the various jaw blank inserts are changed to suit the different workpieces. The lock screw mount hole in the jaws allows the jaws to be held with cap screws for longer production runs where the jaws are not changed as often.

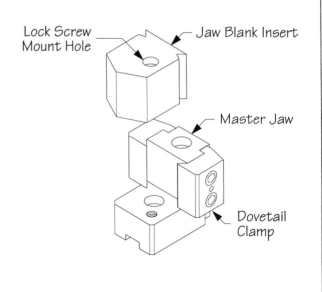

Lock Screw Mount Hole — Jaw Blank Insert — Master Jaw — Dovetail Clamp

# Techniques for Chucks and Collets

**Description of Problem/Requirement:**

Develop a faster, less expensive, and more productive method of changing soft jaws.

**Suggested Solution:**

Use a quick-change mounting system for the soft jaws.

**Source:**

Powerhold, Inc.

**Old Method:**

Although reducing the workpiece mounting time is an important consideration in reducing the total setup time, another consideration is the time required to change the workholding device. Here, using soft jaws also offers several cost savings advantages over building complete workholders. Rather than designing and building a complete fixture for each different workpiece, with soft jaws, only the gripping elements are changed. The actual workholding device—the chuck—remains on the machine tool. This can save a considerable amount of money, since only the jaws are purchased for each job, not the complete workholder. The chuck can be used over and over for many different workpieces.

**New Method:**

One other style of the quick-change soft jaw system uses a horizontal dovetail to position and hold the soft jaws. This design also incorporates two pins in each soft jaw to prevent the jaws from moving when there is no workpiece in the chuck. This system also offers a variety of jaw designs to suit a wide range of workholding situations. In addition to the soft machinable jaws, either smooth or serrated hard jaws are available. The top jaws used with this system are the master jaws that stay mounted to the chuck, while the insert jaws are changed to suit the variety of workpieces mounted in the chuck.

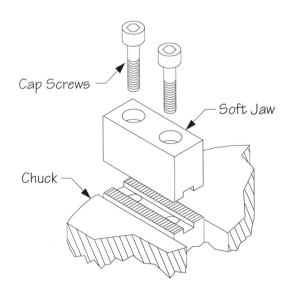

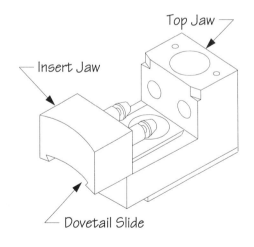

**Description of Problem/Requirement:**

Design an inexpensive alternative to dedicated fixturing for chuck mounted workpieces.

**Suggested Solution:**

Incorporate the Prohold workholding system into the fixturing design.

**Source:**

Enterprise Prohold Company.

**Old Method:**

One method of constructing inexpensive and effective workholders is by combining commercially available universal workholding components. Typically, devices such as chucks and vises may be mounted on angle plates, double angles, tombstones, or many similar components. Individually, these workholders are quite useful; however, when they are combined, they become a low-cost and highly efficient alternative to designing and building dedicated workholders.

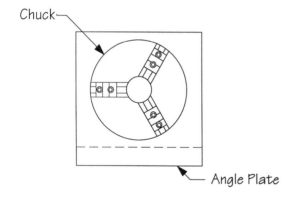

**New Method:**

The Prohold workholding system incorporates a series of chucks built into a variety of multisided tooling blocks. These units are made as two-sided, four-sided, or as single flat pallet models. This design permits up to twelve workstations on a single four-sided fixture, or up to eight workstations on a two-sided unit. All internal components are hardened, ground, and sealed within a fabricated steel body to eliminate any problems caused by dirt and debris.

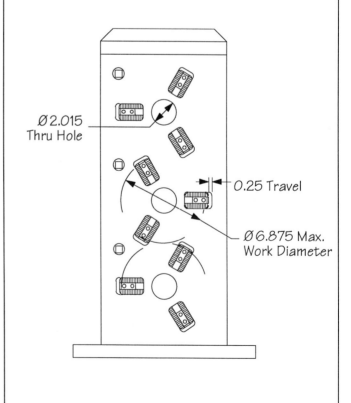

# Techniques for Chucks and Collets

**Description of Problem/Requirement:**

Develop an inexpensive alternative to dedicated fixturing for odd-shaped workpieces.

**Suggested Solution:**

Incorporate the Clamp Chuck workholding system into the workholder design.

**Source:**

Clamp Chuck System.

**Old Method:**

Castings, forgings, and other irregular-shaped workpieces can present a variety of workholding problems. This is especially true when the workpiece must be positioned with respect to an axis, or where multiple features and uneven surfaces must be machined. In most cases, these workpieces require elaborate and expensive dedicated workholders. Specialized, dedicated workholders must be completely designed and built to suit the specific conditions of the workpiece.

Although dedicated workholders may be an ideal choice for high-volume production runs, when only a few workpieces are needed, dedicated workholders are not usually economically justifiable. Here, the workpieces may be mounted on faceplates or clamped in four jaw chucks. Although these accessories are a less expensive alternative to custom fixtures, the setup time required for each workpiece may be quite excessive.

**New Method:**

The Clamp Chuck System may be used for a wide range of turning, milling, boring, and grinding operations. The unit may be spindle mounted or fitted with an optional base unit for table mounted applications. When spindle mounted, the draw tube is attached to the machine draw bar to provide the clamping actions. For machines not equipped with a draw bar, an air cylinder can be attached to the Clamp Chuck to provide the clamping action. When table mounted, the Clamp Chuck unit may be mounted in either a plain mill base or in a hydraulic mill base. When the plain mill base is used, the air cylinder unit may be added for clamping. The hydraulic mill base unit has a self-contained hydraulic clamping cylinder that applies the clamping force.

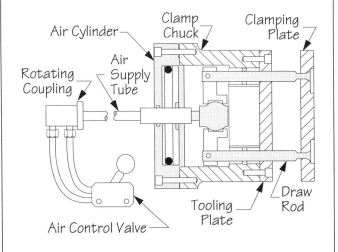

*Description of Problem/Requirement:*

Design a workholder suitable for both prototype and production operations.

*Suggested Solution:*

Design the workholder around the Clamp Chuck workholding system.

*Source:*

Clamp Chuck System.

*Old Method:*

Some workpieces can present a variety of locating and clamping problems. The workpiece shown here is a cast iron bracket. The operations performed in this workholder are drilling, boring, and turning the center hub. Only thirty pieces were required for the initial prototype production run. However, the workpiece had a potential of several thousand pieces per month, once the design was finalized. Here, the main design problem was to create a workholder that would be economical for the initial prototype run, yet capable of handling the full production run. In addition, the workholder design had to be easily modified to suit any design changes.

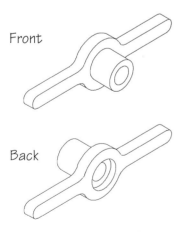

*New Method:*

The workholder was designed using a Clamp Chuck unit. Here, the tooling plate was bored to suit the rough cast hub. In the first operation, the workpiece was loaded with the hub placed in the center hole. The bracket was held in place with a set of finger clamps, operated with the draw rods. The back side of the bracket was then drilled through and the counterbored area was bored to the finished sizes. In the second operation, a locating plug was positioned in the hole in the tooling plate and the workpiece was located on the plug. The drilled hole was then bored to the finished diameter, and the hub was turned and faced to the finished size.

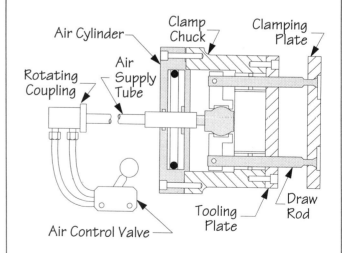

# Techniques for Chucks and Collets

| |
|---|
| *Description of Problem/Requirement:* |
| Design a workholder to hold and locate an odd-shaped workpiece. |
| *Suggested Solution:* |
| Design the workholder around the Clamp Chuck workholding system. |
| *Source:* |
| Clamp Chuck System. |

*Old Method:*

One other example of a problem workpiece is shown here. This workpiece is a cast aluminum pivot block. The operations required for this workholder are turning and facing the diameter on the front side of the block. These machined details are all positioned with respect to the spherical form on the back side of the block. This spherical form and the cylindrical knob at the end were machined in a prior operation. This workholder was designed to verify the proposed processing operations and to act as a prototype for the production workholder. The major setup problem with this workpiece was locating and holding the workpiece with respect to the spherical face.

Front
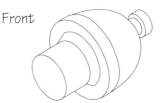

Back

*New Method:*

The prototype workholder was designed and built around the Clamp Chuck System. Here, a cast epoxy block was mounted to the tooling plate to position the spherical form. A special set of draw fingers was added in place of the draw rods. A finger actuation ring was also added behind the tooling plate to close the draw fingers around the cylindrical knob. When activated, the draw tube pulled the draw fingers back. The finger actuation ring then closed these fingers around the cylindrical knob and pulled the block back into the cast spherical recess in the locator. The only change to this initial design that was made for the production workholder was with the cast locator. The final production workholder used a machined aluminum block in place of the cast epoxy block.

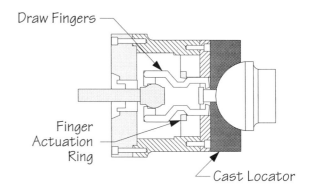

**Description of Problem/Requirement:**

Design a collet arrangement for locating and holding nonstandard workpieces.

**Suggested Solution:**

Incorporate 5C emergency collets into the workholder design.

**Source:**

Various jig and fixture component manufacturers.

**Old Method:**

Collets are frequently used for a variety of workholding setups. Of all the collet variations available today, the 5C is the most common. These collets are available in many standard and special sizes and styles. The most common commercial styles are standard collets, step collets, and expanding collets.

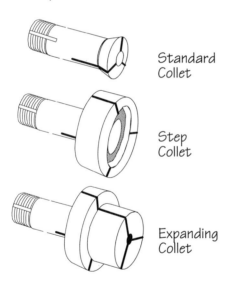

Standard Collet

Step Collet

Expanding Collet

In addition to their wide range of styles and types, another advantage in using 5C collets for workholding is their standardized mounting approach. No matter what the size or shape of the collet, all 5C collets are mounted with their threaded end in a draw bar, draw tube, or nut arrangement. This mounting arrangement allows the 5C collet to be used interchangeably between a variety of machines.

**New Method:**

Although standard-sized collets are widely used, there are other 5C collet variations that should be thoroughly explored as low-cost workholding options. These include 5C emergency collets and step collets. These collets are made with a hardened and ground standard collet mount. The opposite end of the collet is left soft to be machined to suit the requirements of the workpiece. The collets are available in a full range of sizes and styles to suit virtually any requirement.

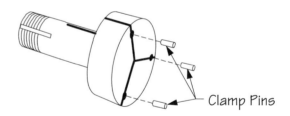

Clamp Pins

The 5C emergency collets and step collets are furnished with clamp pins mounted in holes in the soft end of the collet. These pins hold the collet position during the machining operation. Once the collet is machined, the pins are removed to use the collet. Should the collet require additional machining, the pins are reinstalled to machine the collet. If the collet is intended for a more permanent application, once machined, the collet may also be hardened to reduce wear during the production run.

# Techniques for Chucks and Collets

*Description of Problem/Requirement:*

Design a collet arrangement for locating and holding odd-shaped workpieces.

*Suggested Solution:*

Use a 5C emergency collet machined to suit the workpiece shape.

*Source:*

Various jig and fixture component manufacturers.

*Old Method:*

Standard 5C collets are readily available for a variety of workpiece sizes and shapes. Although the round collet is by far the most popular, collets with both square and hex mounting holes are also commercially available. The standard round collets are available in virtually any size between 0.016" and 1.063". Standard square collets and hex collets are available in fractional sizes to suit standard bar stock sizes.

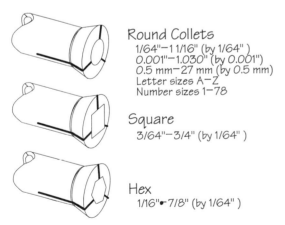

Round Collets
1/64"–1 1/16" (by 1/64")
0.001"–1.030" (by 0.001")
0.5 mm–27 mm (by 0.5 mm)
Letter sizes A–Z
Number sizes 1–78

Square
3/64"–3/4" (by 1/64")

Hex
1/16"–7/8" (by 1/64")

*New Method:*

For either odd-shaped or odd-sized workpieces, the 5C emergency collet may be an ideal alternative to elaborate fixturing. These collets may be machined to suit any workpiece size or configuration within the size range of the collet. Likewise, either standard symmetrical shapes or specialized nonsymmetrical forms may be machined as well.

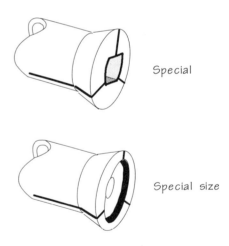

Special

Special size

When machined to suit a particular workpiece, these emergency collets act just like any standard collet and may be transferred from machine to machine to complete the necessary operations. Since the mounting end is hardened and ground, moving these collets between machines only results in a minimal loss in locational accuracy.

*Description of Problem/Requirement:*

Develop a simplified arrangement for indexing collet mounted workpieces.

*Suggested Solution:*

Use standard or customized 5C collet blocks for indexing operations.

*Source:*

Various jig and fixture component manufacturers and in-house fabrication.

*Old Method:*

Both standard and emergency 5C collets are frequently mounted in machine spindles to perform the required operations. However, this is not the only way collets may be used for workholding. A variety of other mounting devices are also available for off-spindle applications. One of the more common devices used for mounting workpieces for off-spindle operations is the 5C indexing head.

The typical 5C indexing head is very useful for a range of workpiece setups. These devices allow a collet mounted workpiece to be indexed for machining flats, slots, holes, or other similar features. The most common indexing units use a direct indexing plate with 24 holes. The 24-hole plate offers the most common numbers of divisions for most workpieces. With this plate, the workpiece can be indexed for 2, 3, 4, 6, 8, or 12 divisions.

*New Method:*

A simpler way of indexing collet mounted workpieces is with 5C collet blocks. These blocks are commercially available with either four or six sides. In use, these blocks are typically mounted in a vise, and index the workpiece by simply rotating the position of the block in the vise.

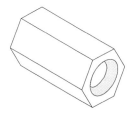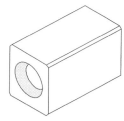

For special operations, a customized collet block can also be made to suit other specialized indexing arrangements. As shown here, two cross-drilled holes, 105° apart, were required in a collet mounted workpiece. A custom made collet block with flats milled at the required angles easily accomplishes this task. Leaving the remaining areas of the collet block with the cylindrical form acts as a foolproofing device to prevent improper loading.

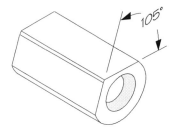

# Techniques for Chucks and Collets

*Description of Problem/Requirement:*

Design a faster, more productive method to switch between chucks and collets.

*Suggested Solution:*

Replace the collet chuck with the Master Jaw System.

*Source:*

EDI Tooling Systems.

*Old Method:*

Increasing productivity involves analyzing the complete production cycle, not just the individual elements. Simply addressing the production volume cannot, by itself, assure better productivity. Both the workpiece and workholder setup must be considered when examining the setup methods. Little is achieved by reducing the workpiece setup time if the workholder setup time is not also improved. Productivity gained in improving the workpiece setup methods can be lost when changing the workholder. One area where this can typically occur is turning.

The two principle types of workholders used for most turning operations are chucks and collets. Chucks are commonly used for larger workpieces, and collets are employed for smaller parts. Changing production jobs often requires switching between standard chucks and collet chucks. Unfortunately, when using conventional equipment, there is no simple or economical way to change these chucks.

*New Method:*

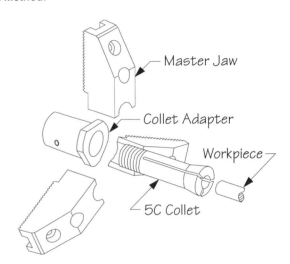

The Master Jaw System transforms a three-jaw power chuck into a 5C collet system. The system consists of three chuck jaws and a 5C collet adapter. This system allows the collets to be installed directly on the chuck, like a set of custom soft jaws, without disturbing the chuck mount or draw tube. The collet is automatically closed around the workpiece as the chuck is tightened. Instead of machining special custom soft jaws for small workpieces, or mounting a collet chuck, the Master Jaw System simplifies the setup by simply switching jaws rather than the complete chuck.

*Description of Problem/Requirement:*

Design a single workholder for multiple operations on the same workpiece.

*Suggested Solution:*

Machine a 5C emergency step collet to suit the workpiece requirements.

*Source:*

Various jig and fixture component manufacturers and in-house fabrication.

*Old Method:*

When using emergency collets for specialized workholding applications, the only limiting factors to consider are the size of the workpiece and the imagination of the designer. Virtually any workpiece that can fit within the size restrictions of the collet can be held. Most often, the emergency collet is machined to suit a single operation on a single workpiece. But, depending on the workpiece, multiple operations using the same collet may also be accomplished.

The workpiece shown here is a cast aluminum connector that requires drilling, reaming, boring, and facing both ends. These machining operations were initially performed with the connector loaded in a three-jaw chuck; however, this method proved to be very slow. When the production volume was increased, a faster method of loading the workpieces was required.

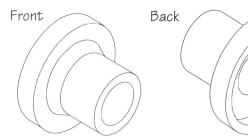

*New Method:*

The design of this workpiece permits a single emergency collet to be machined to suit multiple operations. The first operations are drilling and reaming the center hole and boring the counterbored area. For these operations, the small diameter of the workpiece is mounted in the center hole of the collet. The workpiece is then turned end for end for the second setup. Here the workpiece is held by the large diameter in the grooved face of the collet. In this position, the end of the workpiece is faced to length.

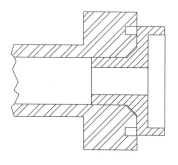

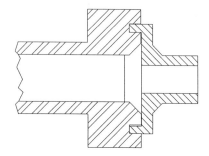

# Techniques for Chucks and Collets

**Description of Problem/Requirement:**

Design a device to set the workpiece to a precise depth within the collet.

**Suggested Solution:**

Add a collet stop to the 5C collet setup.

**Source:**

Various jig and fixture component manufacturers.

**Old Method:**

In addition to the wide array of 5C collets used for workholding, there are several accessory items that should not be overlooked when working with collets. One group of 5C collet accessories that can greatly extend the utility of both standard and emergency collets are the collet stops.

Collet stops are accessories that attach to the mounting end of the collet and allow the workpiece to be positioned to a preset depth in the collet. When using any collet stop, you should remember that the position of the workpiece is relative to the collet and not to the machine tool. So, when the depth of the collet stop is set, it only controls the position of the workpiece with respect to the collet and not the cutting tool.

When tightened, collets have a natural tendency to pull into the spindle. The exact amount of movement is dependent on the size of the workpiece. So, even though the size variation between workpieces may be small, this size variation can have an effect on the actual position of the mounted workpiece with respect to the machine tool.

**New Method:**

### Solid Collet Stop
3.13" maximum chucking depth

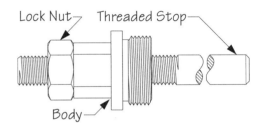

### Long Collet Stop
7.50" maximum chucking depth

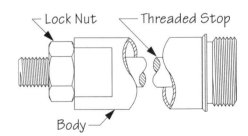

### Spring or Ejector Collet Stop
2.75" maximum chucking depth

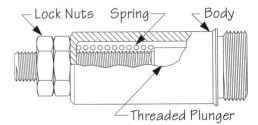

*Description of Problem/Requirement:*

Develop a method of foolproofing the workpiece location within a 5C collet setup.

*Suggested Solution:*

Incorporate a modified 5C collet stop arrangement into the workholder design.

*Source:*

Various jig and fixture component manufacturers and in-house fabrication.

*Old Method:*

The 5C emergency collets are well suited for both primary and secondary machining operations. However, since collets are designed for universal applications, some specialty operations do require a little imagination to accomplish. Foolproofing a collet setup for a symmetrical workpiece is an example of an operation that requires a bit of thought.

Some collet mounted workpieces are relatively easy to foolproof if the external mounting features of the workpieces are different. Here, different collets are often used for each end. However, if both mounting ends of the workpiece are the same and only the internal features of the workpiece are different, other methods must be used to prevent the workpiece from being loaded incorrectly.

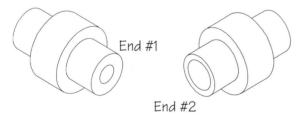

End #1

End #2

The workpiece is a die cast aluminum valve connector with identical external features at both ends. Initially, the workpiece was machined in a standard 5C collet. However, due to a high volume of scrap caused by loading the connectors backwards, a foolproofing device had to be designed to assure proper loading.

*New Method:*

End #2 is loaded first. The diameter of the foolproofing pin prevents loading the workpiece backwards.

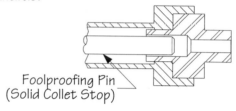

Foolproofing Pin (Solid Collet Stop)

End #1 is loaded next. The workpiece pushes the foolproofing pin into the stop when loaded properly.

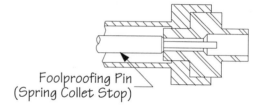

Foolproofing Pin (Spring Collet Stop)

If loaded incorrectly, with end #2 in the collet, the foolproofing pin is not pushed back into the stop. The pin end extends beyond the workpiece to show the operator that the workpiece is loaded incorrectly.

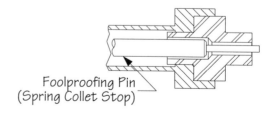

Foolproofing Pin (Spring Collet Stop)

# Techniques for Chucks and Collets

**Description of Problem/Requirement:**

Develop a method to precisely position the workpiece with respect to the machine spindle.

**Suggested Solution:**

Replace the standard collet with a Dead-Length® collet.

**Source:**

Hardinge Brothers, Inc.

## Old Method:

By design, collets are intended to hold parts of a specific diameter. When this diameter varies—even slightly—the collet grips the workpiece differently. Smaller diameter parts will normally pull the collet further into the spindle than will larger diameter parts. This is caused by the fixed size relationship between the collet taper and hole and the variable hole diameter.

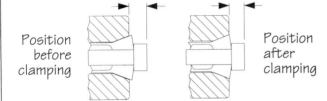

Position before clamping / Position after clamping

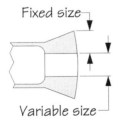

Fixed size / Variable size

This action has little or no effect on the overall accuracy when machining the workpiece to a specific outside diameter. But, when facing shoulders or similar features, to a specific width or thickness, the positional accuracy of the workpiece in the collet is very important.

## New Method:

One answer to this problem is the Dead-Length® collet. This collet design, rather than a one-piece unit, is actually a collet within a collet. The outer collet acts in the conventional manner. But instead of pulling the workpiece into the spindle, the gripping action of the outer collet forces the inner (or Dead-Length®) collet against the face of the spindle nose. This action allows this collet arrangement to achieve extremely accurate positional repeatability from part to part, regardless of any slight variations in the workpiece diameters.

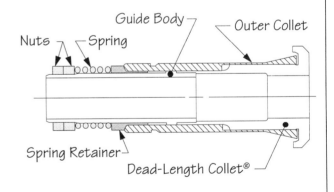

Nuts / Spring / Guide Body / Outer Collet / Spring Retainer / Dead-Length Collet®

# Techniques for Chucks and Collets

*Description of Problem/Requirement:*

Develop a method to precisely position larger workpieces with respect to the machine spindle.

*Suggested Solution:*

Incorporate a modified step chuck and closer arrangement into the workholder design.

*Source:*

Hardinge Brothers, Inc.

*Old Method:*

Length control-type collets are available in a range of sizes and styles to suit a variety of workpiece sizes. However, there are situations where the standard off-the-shelf collets or step chucks cannot perform the operations required. In these situations, the standard collet may still be used, but in a modified form.

The workpiece is a thin-walled ring with a large internal diameter. The width, or thickness, of the ring was machined to a close tolerance. A stepped diameter machined in the bore was also closely controlled. A major concern with machining the ring was the possible deflection or distortion caused when the ring is clamped. For this reason, a collet was determined to be the ideal workholder. However, the thickness tolerance of the ring made using a standard collet setup very time consuming. Each time the workpieces were loaded, they could not be repeatedly clamped in exactly the same position.

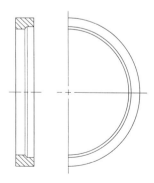

*New Method:*

To accommodate the large diameter of this workpiece, a step chuck is the best choice. However, since the position of the workpiece within the collet is important, a length control device is incorporated into the collet.

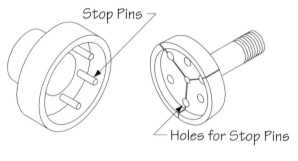

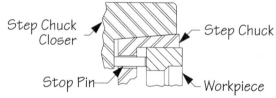

Here, a step chuck and closer were modified to suit the requirements of the workpiece. The step chuck closer is fitted with six pins, and the step chuck has six corresponding holes. The holes in the step chuck are made slightly larger than the pins so the step chuck slides over the pins without interference. When clamped, the precise position of the workpiece is assured by the drawing action of the step chuck. This action pulls the workpiece against the stop pins and establishes a repeatable location from one workpiece to the next.

# Techniques for Chucks and Collets

**Description of Problem/Requirement:**
Design a single collet-type workholder for multiple operations on the same workpiece.

**Suggested Solution:**
Design a simple adapter to accommodate the workpiece in a standard collet setup.

**Source:**
Various jig and fixture component manufacturers and in-house fabrication.

**Old Method:**

Standard, off-the-shelf 5C collet variations are well suited for many machining operations. With only a little imagination, complex or specialty workpieces can be handled as easily as standard operations. Often, a simple adapter is all that is needed to simplify the workpiece processing.

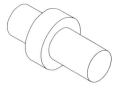 

Rough Cast Blank     Finished Workpiece

This workpiece is a die cast valve body. Both ends of this valve body require machining. The center flange is faced and has an inside radius where the flange meets the short end of the workpiece. The outside diameter of the flange does not require machining and is left in an "as cast" condition.

The original processing of this workpiece was accomplished in a chuck mounted, dedicated turning fixture. However, the main problems with this setup were maintaining the concentricity of the ends to each other and reducing the excessive changeover time between workpieces.

**New Method:**

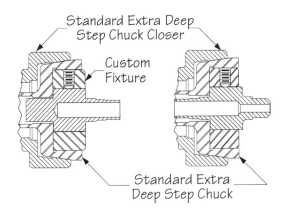

The new fixturing for this workpiece used a standard extra deep step chuck and closer assembly to hold the workpiece in a simple custom adapter. The first operations are facing the flange and turning the three steps on the outside diameter of the long end of the workpiece. Next, the hole is drilled and reamed to a specific depth. The valve body is then flipped end for end and reinstalled in the step chuck. The short end is then turned and faced. The end is chamfered and the smaller hole is drilled into the first hole.

To speed production, two of these custom adapters were used. The operator simply unloaded and reloaded one adapter while the other workpiece was being machined.

### Description of Problem/Requirement:

Design a method to precisely position workpieces with respect to the spindle.

### Suggested Solution:

Incorporate spindle mounted length control devices into the collet setup.

### Source:

Hardinge Brothers, Inc.

### Old Method:

Designers are often very quick to design and build custom dedicated workholders for specialized workpieces. Where practical, a better alternative is to employ existing commercial workholding components. For those cases where off-the-shelf components are not suitable, the wide array of available specialty collets and collet components should be considered. These may be used as is, or customized for special situations.

The Dead-Length® style collet is a very useful specialty collet. But, even these collets have limitations. Occasionally, the size, shape, or other aspects of a workpiece design lend themselves to other types of chucking devices. Here, another type of collet arrangement may be better suited.

The two workpieces shown here are examples where a length control device was needed to act as a fixed locator to precisely machine the required details with respect to their specified reference surfaces.

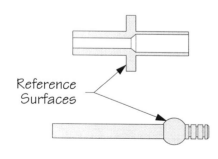

### New Method:

The Dead-Length® Work Stop uses a removeable stop plate as a fixed locator to position the workpiece with respect to the machine spindle. These plates are machined to suit the workpiece requirements. The stop plates are held and aligned in the work stop with conical point set screws. With this setup, the workpiece is held with a standard 5C collet. The drawing action of the collet pulls the workpiece against the stop plate.

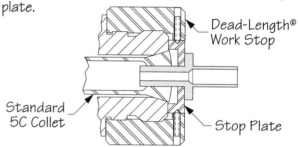

Standard fixture plates may also be used as work stops. The fixture plates are machined to suit the application. Additional elements, such as special locators, may also be added to suit the workpiece. The 5C collet pulls the workpiece into the bushing as the collet is tightened.

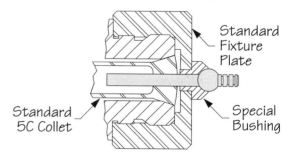

# Techniques for Chucks and Collets

**Description of Problem/Requirement:**

Design a simplified method of holding the workpieces for secondary operations.

**Suggested Solution:**

Incorporate collet accessory devices into the collet setup design.

**Source:**

Hardinge Brothers, Inc.

**Old Method:**

Occasionally, some workpieces are almost impossible to fixture without special-purpose custom dedicated workholders. But, more often, even these workpieces can be fixtured with standard off-the-shelf elements. All that is really required is an understanding of the equipment available and a little imagination.

Secondary machining operations are applications often performed with collet setups. The initial machined details can often make mounting the workpieces simpler. However, sometimes these primary operations can also complicate fixturing the workpieces for the additional operations.

The workpieces here require secondary machining to finish their outside surfaces. The workpiece on the left is also bored and grooved at one end.

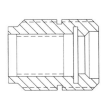
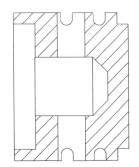

**New Method:**

Workpiece mounted on a threaded arbor and held with a standard 5C collet.

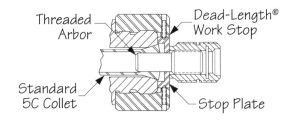

Workpiece mounted on a modified fixture plate and secured with a dowel pin in a modified 5C plug chuck.

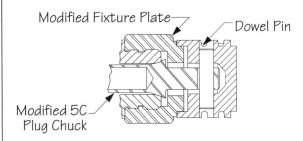

Alternative 5C plug chuck designs for mounting special workpieces.

Screw & Washer

"C" Washer & Groove

*Description of Problem/Requirement:*

Design a simplified method to machine eccentric workpieces.

*Suggested Solution:*

Use modified 5C collets and step chucks to mount the eccentric workpieces.

*Source:*

Various jig and fixture component manufacturers and in-house fabrication.

*Old Method:*

Collet setups are most commonly used for regular-shaped and symmetrical workpieces. However, these are not the only types of machined parts that must be held. Today, odd-shaped and irregular parts are more often the rule rather than the exception. Even many of these complex workpiece designs can be held in collet setups.

One type of specialized part feature that usually requires some creative design alternatives is the eccentric. An eccentric workpiece feature, while not as simple to set up as a straight section, can be machined without too much difficulty in a collet setup. The workpieces shown here illustrate a few setups that may be used to machine these eccentric features.

Each of these workpieces was initially machined with a special dedicated workholder mounted in a four-jaw chuck. While this arrangement made machining the workpieces simpler, the setup time required to install these fixtures in the four-jaw chuck was excessive. Simply replacing the dedicated fixturing elements with modified 5C collets not only produced the workpieces just as fast as the form fixtures, but drastically reduced the setup time.

*New Method:*

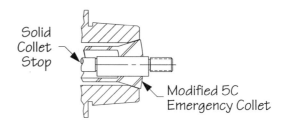

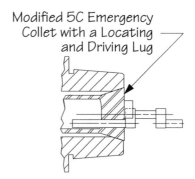

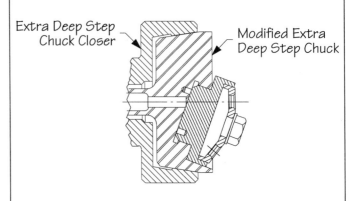

# Techniques for Chucks and Collets

*Description of Problem/Requirement:*

Design a simplified method to machine a larger eccentric workpiece.

*Suggested Solution:*

Design an eccentric adapter for use in a step chuck arrangement.

*Source:*

Various jig and fixture component manufacturers and in-house fabrication.

*Old Method:*

Another eccentric workpiece that presents several workholding problems is shown here. The general shape and size of this part does not lend itself to a standard collet application. The part is an aluminum die casting. The larger outside diameter of the part is turned and the small diameter is left "as cast." The eccentric offset between the diameters is 0.21". With this part, the holes in both ends and the large turned outside diameter are closely controlled.

Initially, these workpieces were turned in a four-jaw chuck because of the eccentric. This method was very time consuming and was quickly abandoned in favor of a faster method. The next attempt at simplification was to use two 5C emergency step collets. Each step collet was machined for each end of the workpiece.

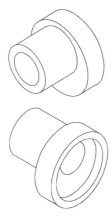

*New Method:*

A better method is to use a standard step chuck, combined with an adapter to suit the eccentric form. This method performs the same operations as before in much less time and at a lower cost.

The adapter is simply two flat plates turned to suit the diameter of the step chuck. The plates are screwed and doweled together. Offsetting the plates 0.21" takes care of the workpiece eccentric. The workpiece is mounted in the adapter on the small diameter and held in place with the clamp screw in the side of the plate.

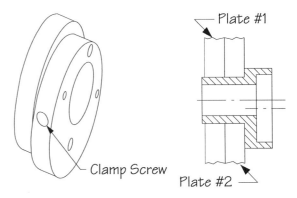

When machined, the adapter is first mounted in the step chuck with Plate #1. When this side is complete, the adapter is reversed and Plate #2 is mounted in the step chuck.

# Techniques for Chucks and Collets

**Description of Problem/Requirement:**

Develop a simplified system to hold workpieces by a center bore.

**Suggested Solution:**

Incorporate a 5C expanding collet into the workholder design.

**Source:**

Various jig and fixture component manufacturers and in-house fabrication.

**Old Method:**

Another collet variation that is very useful for some workpieces is the expanding collet. The basic expanding collet resembles a standard collet at the mounting end. However, at the gripping end, most expanding collets have a series of machinable pads or elements. The specific form of the gripping end varies from one manufacturer to another, but the basic idea and operation of the expanding collet is very similar.

Standard collets grip a workpiece by drawing back into the spindle and collapsing the collet around a workpiece. Expanding collets, however, use the same motion to pull a draw plug or pin through a center hole. This action expands rather than collapses the collet, allowing it to grip the workpiece by an internal surface.

One other style of workholder that is very similar to the expanding collet, but much simpler in design, is the solid screw expanded mandrel. This device is typically a single-piece unit that is mounted in a chuck or collet and turned to suit the workpiece mounting hole. A screw is used to expand the clamping contacts with this unit.

**New Method:**

### 5C Expanding Collet

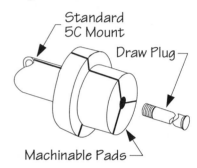

### 5C Master Expanding Collet Assembly

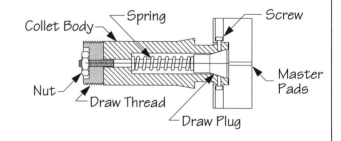

### Solid Screw Expanded Mandrel

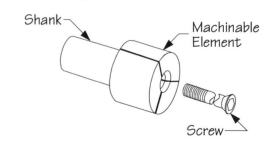

# Techniques for Chucks and Collets

**Description of Problem/Requirement:**

Design an expanding collet arrangement for workpieces with a very small mounting hole.

**Suggested Solution:**

Incorporate a 5C expanding "Mini Collet" into the workholder design.

**Source:**

ROVI Products.

**Old Method:**

The most elementary application of an expanding collet is for workpieces with a simple straight hole. Here, the gripping area of the expanding collet is simply turned to match the diameter of the mounting hole.

The only design limitation with most expanding collets is the size of the workpiece mounting hole. This mounting hole must be made to a specific minimum diameter to allow the draw plug, which expands the collet, to pass through the part to expand the collet. Depending on the design of the expanding collet, the minimum workpiece mounting hole diameter is typically between 0.25" and 0.38". Some expanding collet designs require even larger mounting holes.

Although this may not present a problem with most workpieces, for those situations where a workpiece has a smaller hole size, another type of expanding collet should be used.

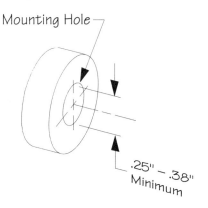

**New Method:**

One specialized type of expanding collet that may be used for mounting holes less than 0.25" is the ROVI 5C Expanding Mini Collet. This collet design is actually a modified form of attachment to a standard expanding collet set. This attachment mounts on a standard mount and allows workpieces with mounting holes between 0.06" and 0.25" in diameter to be mounted on the expanding collet. This particular design recommends that holes larger than 0.25" be mounted on a regular expanding collet.

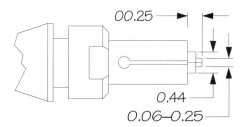

This expanding collet has a 0.44" by 0.25" chucking area for workpiece mounting. Ordinarily, the chucking length of this collet should not be more than 1.5 times the mounting hole diameter. So, if the mounting hole is 0.100" in diameter, the length of the workpiece should not exceed 0.150". The design of this collet applies the holding force by expanding the collet from inside the chucking area rather than with a draw chuck or screw arrangement. This design allows full use of the front part of the expanding collet without interference.

| Description of Problem/Requirement: |
|---|
| Design a simplified method of holding difficult workpieces for machining. |

| Suggested Solution: |
|---|
| Use a combination of a machined expanding collet and a modified step chuck closer. |

| Source: |
|---|
| Various jig and fixture component manufacturers and in-house fabrication. |

**Old Method:**

5C expanding collets will permit a wide range of design possibilities. However, there are times when the expanding collet must be modified or combined with other components to perform special setups.

Front      Back

The workpiece shown here requires machining on both sides. The blank is a bronze casting and it is received with the center hole already drilled and reamed. The problem with this workpiece is supporting the raised and recessed areas on both sides.

Another problem workpiece has a large external diameter and a small mounting hole. The workpiece also requires substantial machining on one side.

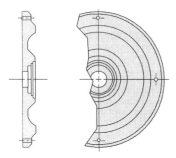

**New Method:**

To resolve this problem, a shoulder is turned on the outer diameter of the expanding collet pads to receive the turned ring. This allows the collet to support the recessed web. An undercut around the expanding element provides clearance for the small turned ring. This assures adequate support for the web when the workpiece is reversed on the collet.

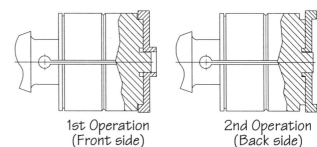

1st Operation      2nd Operation
(Front side)      (Back side)

A modified extra deep step chuck closer and special collar are added to this setup to provide support for the large external diameter. The expanding collet is machined to suit the small mounting hole. A drive pin in the collar prevents the workpiece from slipping during machining.

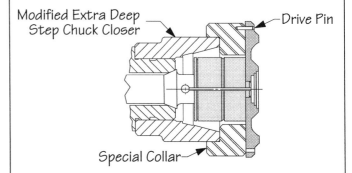

# Techniques for Chucks and Collets

*Description of Problem/Requirement:*

Design a faster method of holding a difficult workpiece for machining.

*Suggested Solution:*

Use a machined expanding collet, modified to suit the workpiece parameters.

*Source:*

Various jig and fixture component manufacturers and in-house fabrication.

*Old Method:*

Many workpieces that formerly required complicated dedicated fixtures are today being machined with less expensive off-the-shelf tooling components and accessories. The workpiece shown here was formerly fixtured in a dedicated fixture and mounted to a faceplate. This method was less than acceptable due to the excessive time the fixture required for loading and unloading the workpieces.

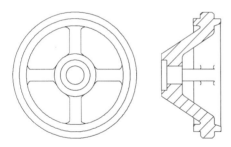

Here, the part is a cast aluminum wheel, approximately 3.25" in diameter. The wheel, as shown, has an offset hub and four spokes connecting the hub to the outer portion of the wheel. Several machining operations are required for the part. These include drilling, reaming, and counterboring the center hole and facing the hub. The outside diameter of the wheel must also be turned to a specific diameter and has two small chamfers cut on the corners. The two shoulders around the outer portion of the wheel are also turned to a specific diameter and faced. All of these operations are performed in a single setup.

*New Method:*

To machine the workpiece in a single setup, the workpiece must be held on the inside diameter of the outer portion of the wheel. Here, a standard expanding collet is modified to suit the workpiece configuration. The diameter of the expanding collet is turned to suit the inside diameter of the workpiece. The front face of the expanding collet has four milled grooves to provide the necessary clearance for the four spokes. Gripping the workpiece on this surface provides both clamping security and sufficient support for the machining.

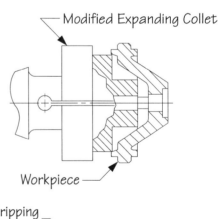

Modified Expanding Collet

Workpiece

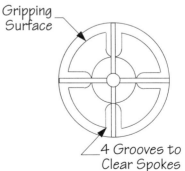

Gripping Surface

4 Grooves to Clear Spokes

## CHAPTER EIGHT

*Techniques for Vises*

# Techniques for Vises

**Description of Problem/Requirement:**
Develop a simplified standardized method of holding workpieces for machining operations.

**Suggested Solution:**
Design the workholding methods around standard and special-purpose vises.

**Source:**
Various jig and fixture component manufacturers.

**Old Method:**

Standard milling machine vises are general-purpose workholders that may be used in a variety of ways. The basic milling machine vise contains both locating and clamping elements. The solid jaw and vise body are the locators. The moveable jaw is the clamping element. The vise is normally positioned so the locators resist the cutting forces. Directing the cutting forces into the solid jaw and vise body ensures the accuracy of the machining operation and prevents workpiece movement. As with all workholders, it is important to direct the cutting forces into the locators. The moveable vise jaw, like any other clamping device, simply holds the position of the workpiece against the locators.

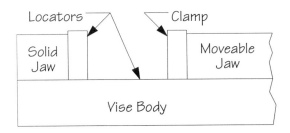

**New Method:**

Although the standard milling machine vises are the most common form of vise used in most shops, there is a wide range of speciality vises that should also be considered, where appropriate, for special-purpose workholding applications. Some of the more general forms of speciality vises are:

- swivel base vises
- angle vises
- compound angle vises
- self-centering vises
- two-piece vises
- hydraulic vises
- pneumatic vises

Regardless of the general form of the vise, all vises operate in much the same way. They all hold the workpiece between a solid and moveable jaw arrangement.

# Techniques for Vises

**Description of Problem/Requirement:**

Design a simplified workholder for positioning workpieces in a vise.

**Suggested Solution:**

Design the workholder around a standard vise with specialized jaws.

**Source:**

Various jig and fixture component manufacturers and in-house fabrication.

## Old Method:

The principle benefits of using a standard vise for holding specialized parts are a self-contained workholding capability, relatively fast lock/unlock cycle, accurate machine mounting capability, fixed position solid jaw, and relatively good rigidity. However, since the standard vise is designed for general-purpose work, specialized workpieces may be difficult to hold. Likewise, production operations which require faster loading speeds are difficult to perform in standard vises due to the lack of any consistent locating points within the standard plain jaws.

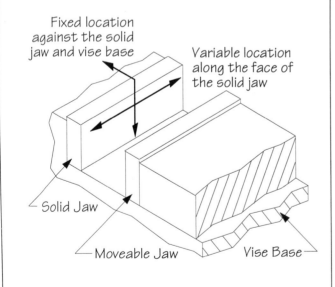

## New Method:

For one-of-a-kind workpieces, or low production applications, this limitation of the standard vise is not a big problem. Here, the vise is used as a simple workholder where the workpiece is mounted between the jaws. However, the standard milling machine vise can be easily transformed into a more specialized workholder with very little effort or expense.

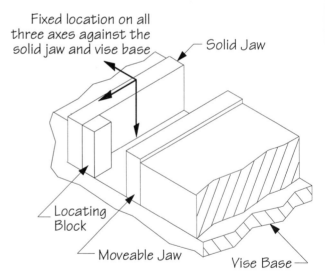

By simply modifying the vise jaws to suit the workpiece, low-cost and efficient workholders can be made at a fraction of the cost of dedicated fixtures. A simple block, added to the solid jaw, acts as a fixed reference for positioning the workpiece in the vise. When analyzing a workpiece for a vise jaw workholder, the workpiece size and shape, the vise size, and the vise jaw material are the only basic considerations.

# Techniques for Vises

**Description of Problem/Requirement:**

Design a vise jaw workholder that assures the positional accuracy of the workpiece.

**Suggested Solution:**

Position the locating elements on the solid jaw and not the moveable jaw of the vise.

**Source:**

In-house fabrication.

## Old Method:

The first step in making a vise jaw workholder is to understand the forces involved in a typical vise mounted machining operation. The cutting (or machining) forces should always be directed into the solid jaw and the base of the vise. This assures better accuracy and safety since these two elements are the most rigid parts of the vise. The cutting forces should never be directed into the moveable jaw.

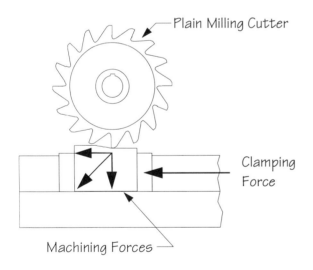

Since the cutting forces should always be directed into the solid jaw, the solid jaw is the element most often modified with a special vise jaw.

## New Method:

For this reason, if a workpiece must be mounted on holes, the solid jaw and not the moveable jaw would be made to suit these locators. In this case, the solid jaw would be made with locating pins to accurately position the workpiece. Since the only function of the moveable jaw is to hold the workpiece against the solid jaw, the moveable jaw is left plain. With all vise jaw workholders, the primary location is accomplished with the solid jaw; the only time the moveable jaw is modified is to support the opposite side of the workpiece and not locate it.

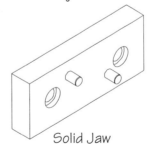
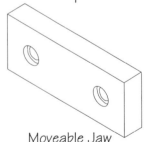

*Description of Problem/Requirement:*

Incorporate mounting elements into a vise setup to hold standard-shaped workpieces.

*Suggested Solution:*

Design a workholder with the locating elements incorporated into the vise jaws.

*Source:*

In-house fabrication.

*Old Method:*

The overall shape and form of the workpiece is an important consideration when selecting the mounting methods. Quite often, additional tooling elements or devices may be added to a vise mounted setup to assist in positioning the workpiece. Two general positioning elements commonly used for these purposes are parallels and "V" blocks.

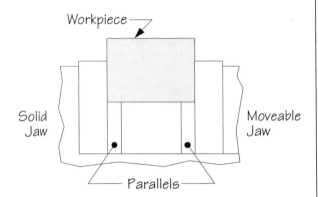

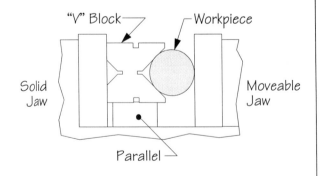

*New Method:*

An alternative to using separate tooling devices to position the workpiece is machining the required shapes into the vise jaws. This method is best suited for workholders that are used repeatedly rather than only one time. Due to the general-purpose nature of both parallels and "V" blocks, these jaws can be used for a variety of different workpieces.

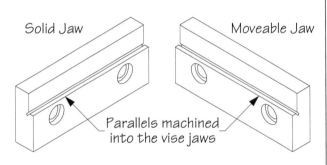

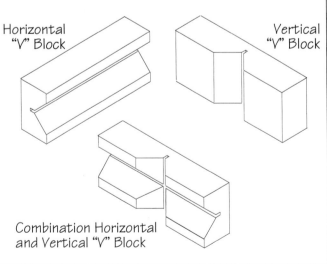

# Techniques for Vises

**Description of Problem/Requirement:**

Design a precise method of mounting and positioning custom vise jaws.

**Suggested Solution:**

Incorporate dowel pin locators or special reference surfaces into the vise jaw design.

**Source:**

In-house fabrication.

**Old Method:**

When designing and constructing special vise jaws, the workpiece is the primary factor in the jaw design. However, if the jaws are expected to be used more than one time, a method of accurately positioning the jaws for additional production runs must be incorporated into the jaw design. One way to do this is with dowel pins inserted into both the solid and moveable jaws. Some vises are made with dowel pin holes machined into both jaws. If a particular vise does not have dowel pin holes, the holes will need to be machined into the vise before this method can be used.

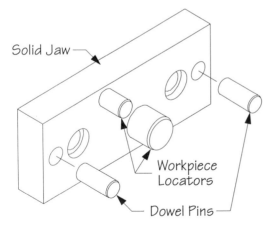

**New Method:**

Installing dowel pin locators, while an excellent method to ensure accurate jaw positioning, is not always practical. In those situations where the dowel pin method cannot be employed, an alternate method of accurately positioning the jaws may be used. Here, the jaws are accurately positioned by simply machining flat surfaces on the custom vise jaws. One flat is machined across the top of the jaw, and a second is machined across the back or front surface of the jaws, at 90° to the top. These surfaces will provide an accurate area for indicating the position of the jaws during the setup.

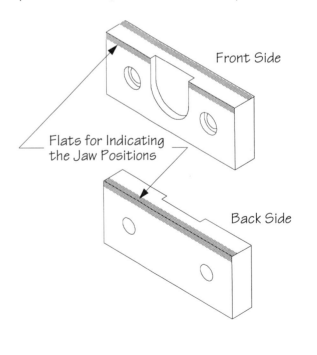

*Description of Problem/Requirement:*

Design a precise method of maintaining the alignment of custom vise jaws.

*Suggested Solution:*

Incorporate alignment pins between the two custom vise jaws.

*Source:*

In-house fabrication.

Old Method:

Although the solid jaw is the primary jaw for positioning the workpiece, there are some workholding situations where both the solid jaw and moveable jaw are customized or machined to properly position and hold some workpieces. Here, the alignment of the two jaws is usually an important consideration. In most cases, the workpiece itself (or machine vise) will maintain the proper alignment of the custom vise jaws.

New Method:

In some workholding situations, the alignment of the solid jaw and moveable jaw is quite critical. Here, two alignment pins may also be installed. These pins will maintain the alignment of the jaws and prevent any lateral movement that might occur with odd-shaped parts or older vises. When installing these alignment pins, always make sure the length of the pins is sufficient to adequately reference the jaws, but not so long as to interfere with the clamping action of the vise.

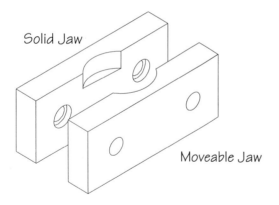

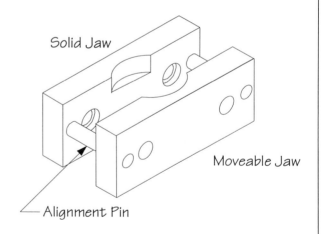

# Techniques for Vises

**Description of Problem/Requirement:**

Design a method of securely holding higher-than-normal workpieces.

**Suggested Solution:**

Use a set of extended height custom vise jaws to suit the workpiece height.

**Source:**

In-house fabrication.

**Old Method:**

Oversize workpieces are another type of problem setup often encountered when using vises. When using a vise, it is usually best to limit the extension of the workpiece above the top of the jaws; this affords the best accuracy and maximum stability, while reducing the possibility of vibration. Although this is the ideal situation, not all workpieces meet these conditions. Here, the choice is often either using an angle plate setup or designing and building a special dedicated workholder.

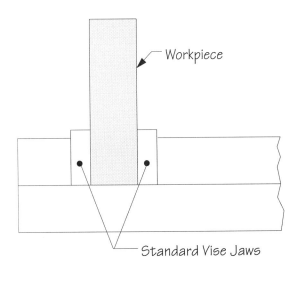

**New Method:**

An alternative method for mounting workpieces that are higher than the normal recommended height is to use a set of custom vise jaws. Here, a set of special jaws may be made to compensate for the additional height of some workpieces. These jaws may be simply made from thicker material if the height is less than 1.50 times the vise depth. For workpieces over 1.50 times the vise depth, additional supports may also be added. When supports are added, some type of adjustment should be designed into the jaw to control any distortion and ensure the accuracy of the setup. When setting up these jaws, always check the setup under a load to make sure there is no distortion or movement in the setup.

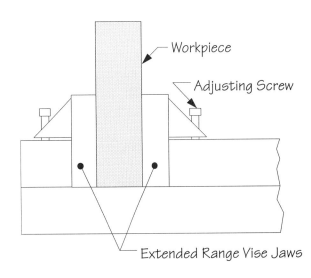

*Description of Problem/Requirement:*

Design a method of securely holding workpieces longer than the normal capacity of a vise.

*Suggested Solution:*

Reposition the custom vise jaws to suit the length of the workpiece.

*Source:*

In-house fabrication.

Old Method:

Extended length workpieces are another form of oversize workpiece frequently encountered when using vises. In any setup using a machine vise, the size of the vise and the size of the workpiece must be compatible. This will help ensure that the vise is able to adequately and safely hold and support the workpiece. With most vise mounted setups, the workpiece is mounted between the solid and moveable jaws. These are the normal mounting surfaces of any vise.

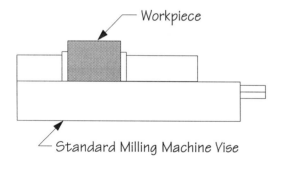

New Method:

An alternative technique for holding longer workpieces is by repositioning the vise jaws. Here, the custom jaws are made higher than the standard jaws, and are positioned to suit the workpiece length. Depending on the specific vise used, these jaws may be able to be positioned on the outer ends of the solid and moveable jaws. When this setup is used, care must be taken not to clamp the workpiece too tightly. When positioned this way, the jaws are acting directly against the screws and not the jaw elements.

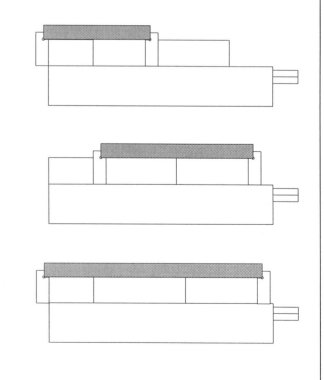

# Techniques for Vises

**Description of Problem/Requirement:**

Design a method of incorporating an adjustable parallel arrangement into a vise setup.

**Suggested Solution:**

Design custom vise jaws with holes to suit the required workpiece heights.

**Source:**

In-house fabrication.

**Old Method:**

Parallels are a common device used in setting up workpieces that must be elevated in a vise. The specific size of the workpiece and the desired height will determine what size parallel should be used. Typically, standard parallels are used for these applications; however, where necessary, special custom parallels may also be used.

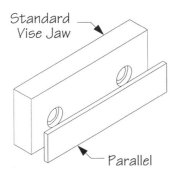

For those situations where a different form of parallel arrangement is necessary, custom vise jaws may be made with the parallels machined directly into the vise jaws.

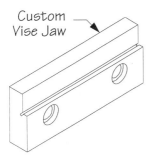

**New Method:**

Custom vise jaws with a series of holes to position a parallel at different heights is an alternative to fixed height parallel arrangements.

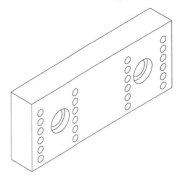

The holes in the face of the jaw are used for 0.25" dowel pins. The dowels can mount the workpiece directly, or act as a support for parallels. The spacing of the holes should be in convenient increments. For general-purpose work, 0.500" spacing is sufficient. Four rows of holes will suit both long and short workpieces. Altering the spacing of one set of holes so they are between the holes in the other set will allow setups in 0.250" increments.

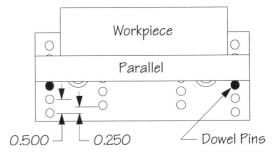

**Description of Problem/Requirement:**

Design a method of securely holding cylindrical workpieces in a standard vise.

**Suggested Solution:**

Construct custom vise jaws with holes along the top of the jaws to mount clamping pins.

**Source:**

In-house fabrication.

**Old Method:**

Vises are unquestionably one of the more popular workholding devices used for manufacturing. However, although vises may be the most desirable general-purpose workholder, there are some workpiece shapes that are not easily set up in standard machine vises.

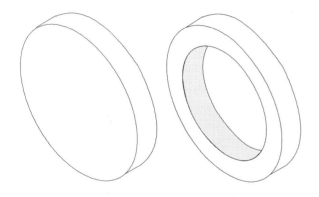

Round or cylindrical workpieces are especially difficult to hold in standard vises. Whether clamped by an external or internal diameter, the two-point clamping contact offered with a regular vise cannot securely hold these workpieces.

**New Method:**

Custom vise jaws with holes on the top surface of the jaws can be used to hold workpieces by both an external or internal diameter. This type of clamping arrangement is preferred for round parts because of the three-point clamping contact.

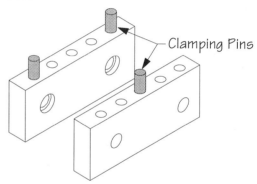

By placing threaded studs in the holes, any large diameter cylindrical workpiece can be securely held. The only difference in the setup is the clamping direction for the external and internal clamping surfaces.

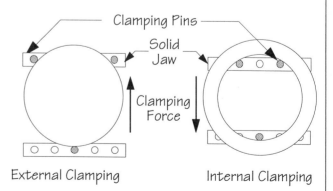

# Techniques for Vises

**Description of Problem/Requirement:**

Design a method of positioning drills and clamping the workpieces vertically in a vise.

**Suggested Solution:**

Design custom vise jaws to mount drill bushings and strap clamps.

**Source:**

In-house fabrication.

**Old Method:**

Drilling is an operation frequently performed in a vise. Although a vise may be well suited for holding a workpiece, holding is all a standard vise can do. There is no provision in a standard vise for production drilling operations. So, while the workpiece may be held, some other device must be used to precisely position the drill relative to the workpiece.

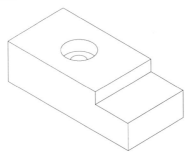

Another tricky setup in a vise is holding multiple workpieces. Usually, the workpieces are stacked in the vise horizontally, so the clamping action of the vise forces them together. However, when the workpieces must be stacked vertically, a standard vise cannot apply a vertical clamping force to hold the workpieces together.

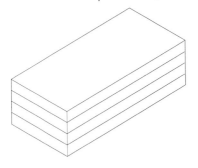

**New Method:**

Placing a series of tapped holes across the top of the vise jaw permits a variety of attachments to be mounted to the vise. For drilling, one or more bushing arms could be mounted to the jaw. Placing liner bushings in the bushing arms allows different size drill bushings to be used.

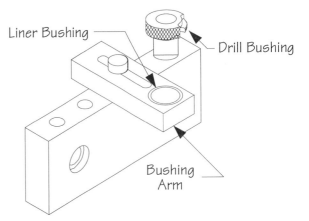

The bushing arms may also be replaced with strap clamps for those applications where the vise must be used to apply vertical clamping forces.

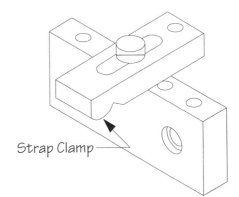

*Description of Problem/Requirement:*

Design a universal method of locating different size workpieces parallel to the jaws in a vise.

*Suggested Solution:*

Incorporate an adjustable work stop into the end of the vise jaw.

*Source:*

In-house fabrication.

*Old Method:*

The design of the typical machine vise will permit a workpiece to be located and held on two of its three axes. These two are the vertical axis and the horizontal axis, perpendicular to the jaws. The third axis, which is also horizontal, but parallel to the vise jaws, is usually unrestricted. To accurately position a workpiece, all three of these axes must be completely restricted.

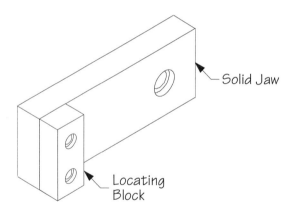

One method frequently used to restrict the horizontal axis that is parallel to the jaws is to add a locating block to the solid jaw. While quite adequate, a fixed block will not permit any adjustment to accommodate different workpieces. Likewise, since the block is often mounted on the face of the solid jaw, longer workpieces cannot be easily positioned.

*New Method:*

A more universal alternative to the fixed locating block is an adjustable work stop, mounted to the solid jaw of the vise. Here, a series of tapped holes is added to the end of the vise jaw. A mounting stud is then inserted into one of these holes. A variety of different length studs could be used to suit different length workpieces. An adjustable work stop is then added to the mounting stud and positioned to suit the location of the workpiece in the vise. Coarse adjustments are made by sliding the complete adjustable stop along the mounting stud. Fine adjustments against the workpiece are made by tightening or loosening an adjusting screw in the end of the work stop. This design also permits the work stop to pivot to suit any workpiece thickness.

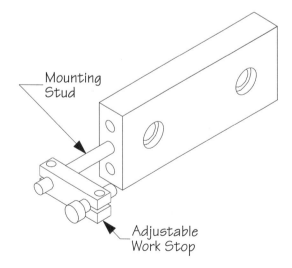

# Techniques for Vises

| |
|---|
| *Description of Problem/Requirement:* <br> Develop a method of horizontally clamping multiple workpieces in a vise. |
| *Suggested Solution:* <br> Add an end stop and a side clamping device to the solid vise jaw. |
| *Source:* <br> In-house fabrication. |

*Old Method:*

Clamping multiple workpieces in a vise can be a tricky operation. No matter how carefully the workpieces are mounted, the vise will only clamp against the two largest workpieces. The other workpieces are actually floating free between the vise jaws. For this reason, workpieces often wobble or move during machining.

One way to avoid this problem is to place the larger workpieces at the outer ends and hope they will retain the other workpieces. Another method is to add an auxiliary clamping device to the setup.

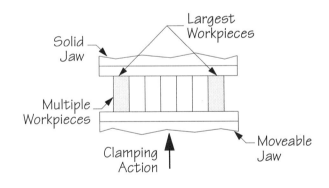

*New Method:*

An alternative method of setting up multiple workpieces in a vise is with a side clamping device incorporated into the design of the solid jaw. This device includes an end stop and a side clamp, mounted to the solid jaw.

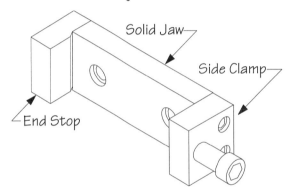

When used, the side clamping device adds a side clamping action to the setup. This allows the vise to clamp the workpieces in one direction and the side clamp to apply a clamping force perpendicular to the clamping direction of the vise. This will securely hold all the workpieces, regardless of minor variations in their size.

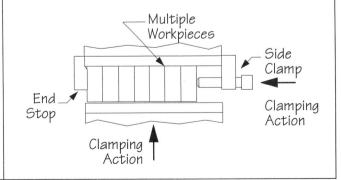

### Description of Problem/Requirement:
Design a method to accurately position the cutter and securely hold thin workpieces in a vise.

### Suggested Solution:
Make a set of vise jaws to support the workpiece with a cutter setting device in the solid jaw.

### Source:
In-house fabrication.

### Old Method:

Accurately locating the cutting tool with respect to the workpiece is often a problem with vise setups. Standard vises only hold the workpiece and have no provisions for locating. Another difficulty in working with vises is mounting thin workpieces. This is especially true when machining details such as slots. Quite often, the workpiece lacks sufficient rigidity to permit the workpiece to be machined without excessive chatter and vibration.

The workpiece shown here is a thin steel bar requiring a milled slot. The workpiece must be set up to reduce any vibration caused by the milling operation. The relationship between the cutter and the workpiece may be established by cut-and-try methods on the first workpiece. Subsequent workpieces are then machined with these settings. This method is suitable for a few workpieces but may be too time consuming for higher production volumes. To reduce any vibration with this thin workpiece, a smaller cutter is used and the slot is cut in two passes.

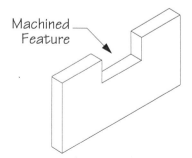

### New Method:

A set of custom vise jaws is made to machine this thin workpiece. These jaws are made to both support the workpiece and to provide cutter referencing. The slot in the solid jaw acts as a set block, and the cutter position is set with a thickness gauge. The first pass is referenced against the bottom and one side of the slot, while the second pass is positioned with the opposite side of the slot. The slot in the moveable jaw is made slightly larger to provide the necessary cutter clearance.

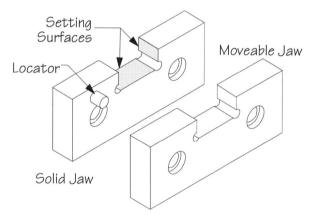

The slot is larger in the moveable jaw to provide cutter clearance

# Techniques for Vises

**Description of Problem/Requirement:**

Design a faster method of changing vise jaws on standard machine vises.

**Suggested Solution:**

Replace the standard vise jaws with the SNAP JAWS® Quick-Change Vise Jaw System.

**Source:**

Turnmill Machine Company.

**Old Method:**

Combining special-purpose or custom vise jaws with standard machine vises is a cost-effective alternative to building dedicated workholders. However, with most standard machine vises, the time required to change the jaws to suit different jobs is often quite excessive. This is due in large part to the methods used to attach both standard and custom jaws. Conventional machine vise designs typically use socket head cap screws to mount the jaws. Although a common method, this design requires both mounting screws be completely removed to change the vise jaws.

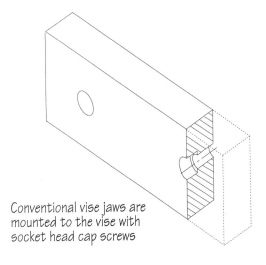

Conventional vise jaws are mounted to the vise with socket head cap screws

**New Method:**

The SNAP JAWS® System is an extremely rapid, yet accurate and secure, jaw mounting system. Standard low-head cap screws and a "T"-slot arrangement are used to mount the jaws. The screws are installed in the mounting surfaces of the fixed and moveable vise jaws. The jaws are installed by aligning the "T"-slots in the jaws with the screws and lowering them into position. The jaws are secured by tightening the screws through the holes in the jaws. The complete changeover takes approximately 10 seconds.

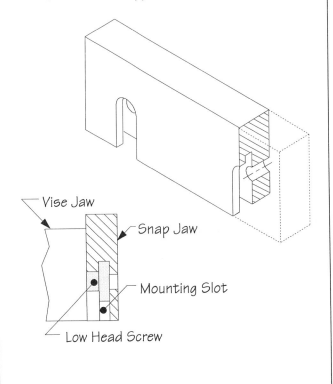

| |
|---|
| **Description of Problem/Requirement:** |
| Design a method of securely holding odd-shaped workpieces in a standard vise. |
| **Suggested Solution:** |
| Position a Force 14 vise accessory between the workpiece and the vise jaw. |
| **Source:** |
| Kopal Products. |

**Old Method:**

The workpieces held in most vises generally have a regular shape that is easy for the vise jaws to grip. However, not all workpieces are designed and constructed to be easily mounted in a vise. Odd shapes, stepped shoulders, and similar workpiece features can often make mounting and securely holding the workpiece difficult.

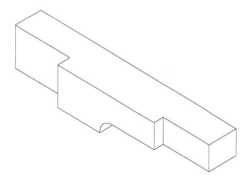

The workpiece shown here has stepped portions on both ends. These limit the contact area of the workpiece against the vise jaw. The workpiece could be held between two plain vise jaws. However, the additional clamping force required to securely hold the workpiece during the machining could damage or deform the workpiece.

**New Method:**

The Force 14 vise accessory incorporates a series of self-positioning and compensating pistons into a block that is positioned between the workpiece and the vise jaw. When the vise is closed, these pistons contact the workpiece and hydraulically compensate for different surface positions. The hydraulic reservoir is contained within the block, and as one piston is pushed in, the others are moved out.

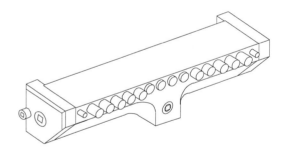

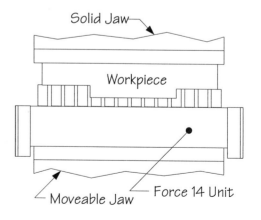

# Techniques for Vises

*Description of Problem/Requirement:*

Design a method of securely holding odd-shaped workpieces in a standard vise.

*Suggested Solution:*

Position a Swivel Jaw vise accessory between the workpiece and the moveable jaw of the vise.

*Source:*

Tayco Tools, Inc.

*Old Method:*

Odd- and irregular-shaped workpieces can really be a problem to hold securely in a vise. Workpieces with compound angles are especially difficult to fixture with a vise. Here, there is no simple way to match the workpiece shape to the vise jaw. Although a dedicated fixture might be one alternative, when just a few workpieces are being made, it is very difficult to justify the added cost of the fixturing.

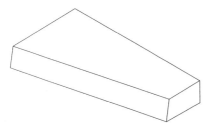

With this workpiece, the straight surface is intended to be the reference surface, and the clamp is to be directed against the surface with the compound angle.

*New Method:*

One way to simplify holding this workpiece is with the Swivel Jaw vise accessory. This unit is a rectangular block housing a swivel unit. This block provides a stable mount, and locates and aligns the accessory in the vise. The main swivel unit has one flat side and a semicircular form on the opposite side. The semicircular shape moves, in an arc, within the mounting block. Two flattened balls are contained within the swivel element. These balls swivel in their sockets. This permits each ball to move independently of the other, and allows the Swivel Jaw to conform to irregular angular or tapered forms with either simple or compound angles.

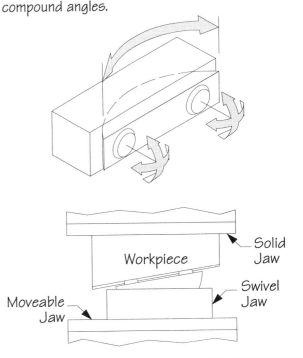

**Description of Problem/Requirement:**

Develop an alternate method of holding two workpieces in a standard machine vise.

**Suggested Solution:**

Install the Twin Lock vise accessory on a standard machine vise.

**Source:**

Discount Tool.

**Old Method:**

The most common forms of machine vises are designed with one fixed jaw and one moveable jaw. This standard vise design, while quite adequate for many applications, does limit the overall functionality and utility of the machine vise. Many attachments and accessories are available for vises to overcome some of the inherent design limitations. But in spite of these devices, the basic two-jaw design of the machine vise remains a serious limiting factor in fixturing a variety of specialized workpieces with machine vises.

**New Method:**

The Twin Lock vise accessory offers a slightly different approach to holding workpieces. This accessory consists of two jaw blocks which replace the moveable jaw on any standard angle locking-type vise. These jaw blocks will hold single or multiple parts and have the unique capability of applying both horizontal and vertical hold force. The jaws are mounted on a yoke-and-piston arrangement, and can compensate for workpiece variations up to 0.500". This allows two different size parts to be held and clamped with the same holding force. The pistons used to mount the jaws have vertical travel of 0.100".

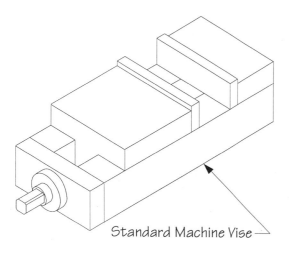

Standard Machine Vise

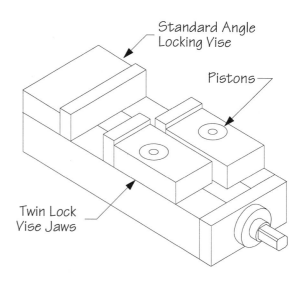

Standard Angle Locking Vise

Pistons

Twin Lock Vise Jaws

# Techniques for Vises

**Description of Problem/Requirement:**
Develop an alternate method of holding special workpieces in a standard machine vise.

**Suggested Solution:**
Install the Twin Lock vise accessory on a standard machine vise.

**Source:**
Discount Tool.

**Old Method:**

Performing special setups in standard machine vises can sometimes be a problem. This is especially true when special shapes or multiple workpieces must be held. Two workpieces may be held in a standard machine vise without too much difficulty, if they are the same size. However, two different size workpieces, or a stepped workpiece, are much more difficult to set up with a standard vise. The basic design of the standard machine vise does not lend itself well to holding specialized workpieces without special-purpose (and often expensive) fixturing elements.

**New Method:**

The dual clamping jaw design of the Twin Lock vise accessory allows different size workpieces and stepped workpieces to be clamped as easily as two identical workpieces. The design of this attachment automatically adjusts to accommodate the different sizes of two workpieces held side by side. Likewise, when a single workpiece is clamped, the side of the opposite jaw may also be used to assure that the workpiece is properly located and positioned at 90° to the machine spindle.

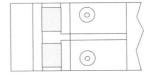
2 Identical Workpieces

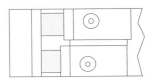
2 Different Workpieces

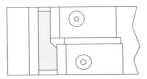
Stepped Workpiece

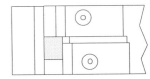
Single Workpiece at 90°

*Description of Problem/Requirement:*

Develop an alternate method of holding special workpieces in a standard machine vise.

*Suggested Solution:*

Install the Twin Lock vise accessory on a standard machine vise.

*Source:*

Discount Tool.

*Old Method:*

Other types of special vise setups that are often quite difficult involve clamping angular, contoured, or similar irregular workpiece shapes. Likewise, holding oversize workpieces or mounting multiple workpieces are other tasks not normally performed in standard machine vises. Once again, the basic design of the standard machine vise often limits its application to more standardized (or regular) workpiece shapes. So, although a vise may be the ideal workholder for some jobs, the size, shape, or complexity of the workpiece features will often prevent standard machine vises from being employed.

*New Method:*

For some special shapes, the clamping elements may be mounted to the pistons. Here, a set of swivel jaws is attached to the pistons to hold a variety of odd workpiece shapes. These swivel jaws are usually customized to suit the part shape. For oversize workpieces, a set of strap clamps may also be mounted to the pistons. The clamping action of the vise pulls the pistons down, locking the strap clamps. Multiple small workpieces can also be clamped in a single setup using a set of special jaws and an anvil plate. As the vise is tightened, the swivel jaws contact the anvil plate and push it against the parts. As the pressure increases, the swivel jaws pivot and apply the necessary holding force to both ends of the stacked parts.

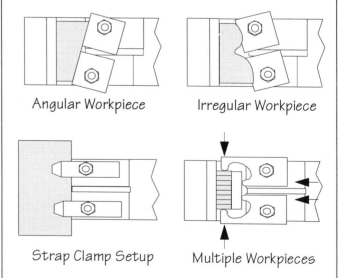

Angular Workpiece    Irregular Workpiece

Strap Clamp Setup    Multiple Workpieces

# Techniques for Vises

**Description of Problem/Requirement:**

Incorporate a lighter and more rigid vise design into the workpiece setup.

**Suggested Solution:**

Install the Quad-1 Precision Vise into the workholding arrangement.

**Source:**

Interlen Products Corporation.

**Old Method:**

The machine vises commonly used today, for all their benefits, do occasionally have some serious drawbacks. Most notable is the basic clamping action of the standard vise. Most vises used today use a pushing action to hold a workpiece between a moveable and a solid jaw. This push-type clamping arrangement does have a few inherent design flaws. The most common problem in using a push-type vise arrangement is the tendency of the moveable jaw to raise as the clamping pressure is applied. Likewise, the vise screw used with this vise design also has a tendency to buckle or distort as the torque on the screw is increased. To overcome these difficulties, many standard machine vises use added bulk and mass to achieve the necessary strength and rigidity.

**New Method:**

One method to eliminate, or at least reduce, these problems is to use a clamping arrangement that pulls the moveable jaw, rather than pushes it against the workpiece. This pull-type arrangement permits the vise to develop significant torque while reducing the distortion of the vise screw. Likewise, the construction of the moveable jaw reduces or eliminates the tendency for it to raise when clamped. The basic design and general construction of this vise also helps to maintain impressive accuracy, while reducing the weight. Rather than using added mass to assure the necessary strength and rigidity, this vise relies on its design and construction to achieve and maintain its accuracy.

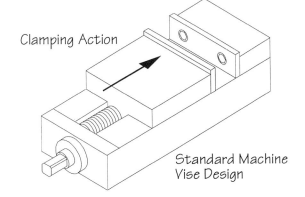

Standard Machine Vise Design

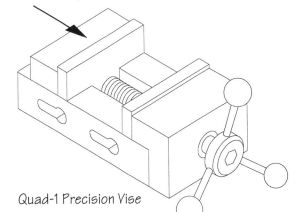

Quad-1 Precision Vise

| |
|---|
| *Description of Problem/Requirement:* |
| Devise a standard vise capable of holding various odd-shaped workpieces. |
| *Suggested Solution:* |
| Incorporate the Multi-Fixture Vise into the workholder setup. |
| *Source:* |
| James Morton Company. |

*Old Method:*

The basic idea behind the standard machine vise has always been to use a general-purpose workholder to accurately and securely hold a variety of specialized workpieces. Typically, this function is accomplished by holding the workpiece between two flat jaws. However, despite the overall utility of this design, there are workpiece shapes that simply cannot be held securely between two flat surfaces.

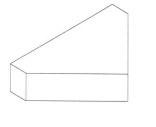

Some typical problem parts that cannot be securely held in a plain machine vise are cylindrical, triangular, or other odd-shaped parts. These workpiece shapes often require special fixturing, or at least a different workholding method. Occasionally, a three-jaw or four-jaw chuck may be used to hold some workpieces. But the lack of a fixed reference point with these chucks, as well as the problems of mounting a chuck on a machine table, often restricts their use for these purposes.

*New Method:*

A vise that combines the basic functions of a vise, three-point contact of a chuck, and infinitely adjustable jaws is the Multi-Fixture vise. This vise has three jaws that may be set to conform to the workpiece. These jaws are made from layers of stacked plates. This permits the jaws to be moved to form individual shelves, much like parallels. Once positioned, the shafts on the side of the vise clamp the jaws with a cam locking mechanism. Once locked, the vise is operated like any other vise. The jaws retain their original setting until unlocked and reclamped for another part shape.

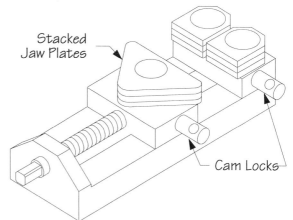

The design of this vise permits a variety of odd-shaped workpieces to be held very easily.

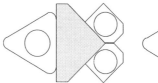

# Techniques for Vises

**Description of Problem/Requirement:**

Develop a method of mounting two workpieces in a single machine vise.

**Suggested Solution:**

Incorporate a Chick Qwik-Lok™ vise into the workholding arrangement.

**Source:**

Chick Machine Tool, Inc.

**Old Method:**

Plain milling machine vises, while well suited for a variety of general-purpose workholding operations, lack many of the features needed for most production machining operations. Production machining operations typically require faster and fewer setups to be made to maintain the overall speed and efficiency of the operation. Standard vise designs typically only permit single workpieces to be held in any single setup. Although special jaws may allow two workpieces to be mounted side by side, the size of the vise is often a limiting factor in fixturing larger workpieces.

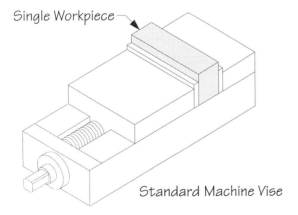

Standard Machine Vise

**New Method:**

An alternative design that can help increase production by doubling the capacity of the vise is the Chick Qwik-Lok™ Vise. This vise has two separate gripping positions. Either two identical or two totally different sized workpieces can be held. The clamping mechanism of the vise is totally self-compensating and will securely clamp both workpieces. The same gripping pressure is obtained on both gripping positions regardless of the workpiece sizes.

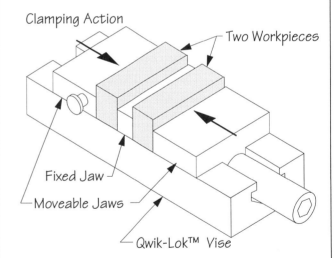

The center jaw on this vise is the fixed jaw. The two outer jaws are the moveable jaws, and they both clamp inward, toward the fixed jaw. For holding very large single workpieces, the center jaw may be removed and the workpiece clamped between the two outer jaws. The design of this vise uses a pull-type jaw action that eliminates workpiece deflection and lift.

## Techniques for Vises

*Description of Problem/Requirement:*

Develop a method of vise mounting several workpieces in a single machine setup.

*Suggested Solution:*

Incorporate a Chick Multi-Lok™ vise fixture into the workholding arrangement.

*Source:*

Chick Machine Tool, Inc.

*Old Method:*

With today's emphasis on higher production volumes, more machine tools are being set up with workholding devices capable of holding several workpieces. When vises are used to hold these workpieces, one of the more popular methods of mounting the vises is with a multisided cube, or tombstone. Although very useful, with some machining centers, the machining envelope is too small to permit large tombstones to be used. So, occasionally a two-sided double angle is used in place of a four-sided tombstone to reduce the overall size.

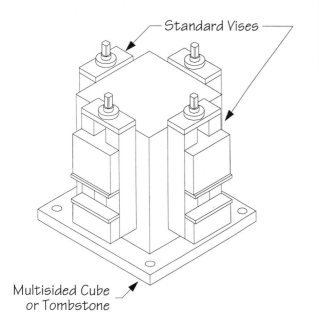

Standard Vises

Multisided Cube or Tombstone

*New Method:*

Another form of vise fixture that is very well suited for multiple workpiece setups is the Multi-Lok™ vise fixture. This unit incorporates four Qwik-Lok™ vises into a single tombstone-like unit. Unlike mounting standard vises to a tombstone, these vise units are built into the column and require much less space on the machining center table. So, even a machine with a small envelope can use these units. Each vise unit will hold two workpieces for a total of eight for the complete fixture unit. The design of these vises also permits eight identical or eight completely different workpieces to be mounted in any setup.

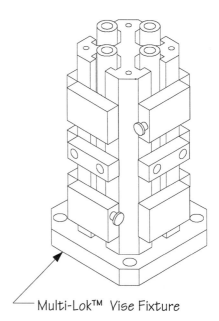

Multi-Lok™ Vise Fixture

# Techniques for Vises

**Description of Problem/Requirement:**

Incorporate a method of mounting quick-change custom jaws for vise setups.

**Suggested Solution:**

Design the vise fixture around the QwikChange™ Machinable Jaw arrangement.

**Source:**

Chick Machine Tool, Inc.

**Old Method:**

With most conventional machine vises, both the standard and custom jaws are attached with socket head cap screws. To change these jaws, the screws are usually removed completely, the new jaws are then positioned in the vise, and the screws are reinstalled. Although a relatively simple process, the time required to change the jaws is often quite excessive.

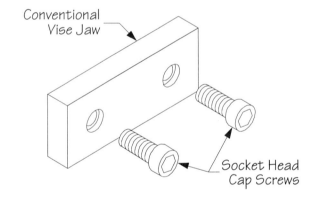

**New Method:**

Both the Qwik-Lok™ vise and the Multi-Lok™ vise fixture use a rather unique jaw mounting system. This system uses a single "D"-shaped pin inserted through each jaw into a "D"-shaped hole in the jaw carrier. This arrangement applies both lateral and downward force on the jaws as they are tightened. The QwikChange™ Machinable Jaws are designed to be machined to suit the specific shape of the workpiece.

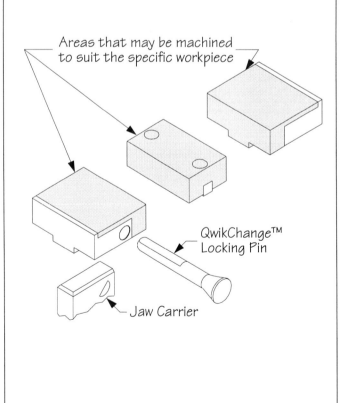

## Techniques for Vises

**Description of Problem/Requirement:**

Devise an efficient way to mount workpieces vertically using a vise setup.

**Suggested Solution:**

Design the workholder setup using machining center vises.

**Source:**

Dapra Corporation.

**Old Method:**

Standard machine vises may be mounted in a variety of ways to suit a range of applications. Double angles, or "T" fixtures, are one device frequently used to mount vises for vertical setups. The vises may be mounted on one or both sides of the double angle, depending on the requirements and the machining envelope of the machine tool. The time required to mount the vises is a major drawback to this type of setup. Likewise, the added size of the double angle may also limit the sizes of workpieces that may be mounted.

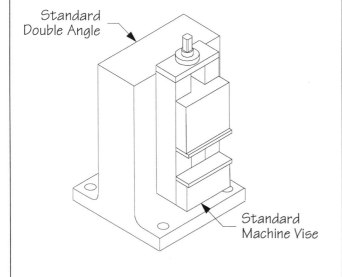

Standard Double Angle
Standard Machine Vise

**New Method:**

The machining center vise is an ideal workholding alternative for setups on machines with small envelopes. These vises may be used individually, or mounted back to back. A unique spindle design allows the vises to be used for both internal and external clamping arrangements. The vise also has a mechanically intensified high pressure spindle for precise clamping pressures. The box column design of the vise base unit makes the vise extremely rigid to assure consistent accuracy. Several attachments are available to permit the vise to perform a wide range of different setups.

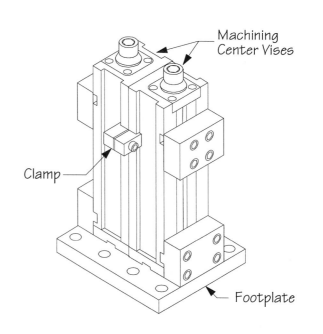

Machining Center Vises
Clamp
Footplate

# Techniques for Vises

**Description of Problem/Requirement:**

Design an efficient vise setup for mounting very large workpieces directly to a machine table.

**Suggested Solution:**

Design the workholder setup using a Maxi-Mill Machine Vise.

**Source:**

Tayco Tools, Inc.

**Old Method:**

One serious shortcoming common to many standard machine vises is their capacity. Although some vises are available with larger openings, most are limited to between 6" and 8". This is adequate for most shop tasks, but occasionally the size of the workpiece may be beyond the limits of the available vises. So, in these setups, even though a vise would be the ideal workholder, other workholding devices must be used. Here, strap clamps are often used simply because the size of the workpiece is too large.

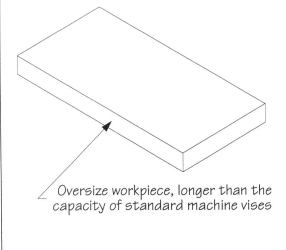

Oversize workpiece, longer than the capacity of standard machine vises

**New Method:**

The Maxi-Mill Vise solves the problem of holding oversize workpieces. This unit is a two-piece vise arrangement. The solid (or stationary) jaw is attached to one end of the machine table, and the moveable (or clamping) jaw is mounted at the other end. This arrangement permits the jaws to be spaced at any distance within the size of the machine table. With this vise, virtually any part that can fit on the machine table can be held. A rack gear, positioned in the center "T" slot in the table, acts as a tie bar to hold the jaw positions. This tie bar prevents the jaws from moving apart when the vise is tightened against the workpiece.

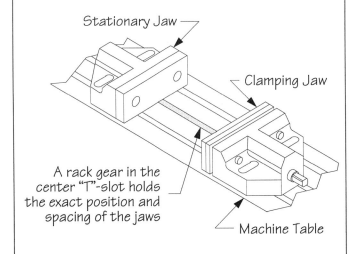

Stationary Jaw

Clamping Jaw

A rack gear in the center "T"-slot holds the exact position and spacing of the jaws

Machine Table

## Techniques for Vises

**Description of Problem/Requirement:**

Devise an efficient vise arrangement to hold very large workpieces.

**Suggested Solution:**

Incorporate a two-piece modular vise into the workholder design.

**Source:**

Hilma Corporation / Carr Lane Manufacturing Company / Stevens Engineering.

### Old Method:

Some oversize workpieces are quite difficult to fixture. This is especially true when working with machine tools with small tables, or when using pallets. Here, the workpiece must often be clamped with a variety of special-purpose clamping devices, or expensive custom designed fixtures. Many times, simpler workholders would be preferred, but because of the overall size or shape of the workpieces, these standard workholders cannot be used.

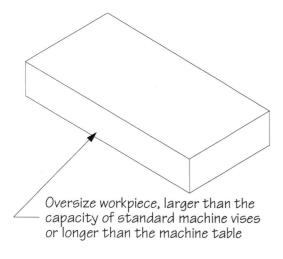

Oversize workpiece, larger than the capacity of standard machine vises or longer than the machine table

### New Method:

One answer to the oversize workpiece problem is to use a two-piece modular vise. The elements may be positioned on a modular subplate or mounted on a custom baseplate. With this design, only the size of the baseplate unit limits the capacity of the vise. For larger workpieces, the vises may also be ganged together. In addition to the basic clamping units, there is a wide range of speciality clamping jaws and devices available as accessories for this vise.

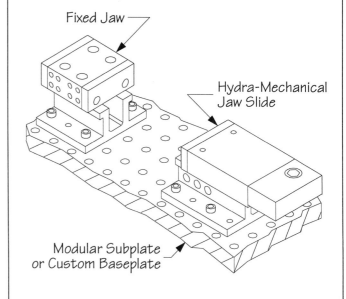

Fixed Jaw

Hydra-Mechanical Jaw Slide

Modular Subplate or Custom Baseplate

# Techniques for Vises

**Description of Problem/Requirement:**

Incorporate a self-centering clamping arrangement for holding round workpieces.

**Suggested Solution:**

Design the workholding arrangement around a self-centering vise.

**Source:**

Carr Lane Manufacturing Company.

**Old Method:**

A cylindrical object is a workpiece shape that can be tricky to hold. This is especially true when the setup must be self-centering or aligned to the workpiece centerline. One of the most common ways to mount these workpieces is with a "V" block. Once positioned in the "V" block, a strap clamp is often used to hold the workpiece.

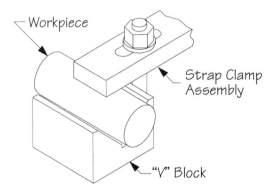

**New Method:**

Another way to hold cylindrical workpieces is with a self-centering vise. These vises are available to accommodate a wide range of workpiece sizes. Most of the better self-centering hydraulic vises operate with a rack-and-pinion mechanism and can repeat their position from part to part to within 0.002", when the same pressure is applied. These vises may also be furnished with different jaw sets to suit different workpiece diameters. The model shown here will hold workpieces between 0.25" and 2.94". Larger self-centering vise models will hold workpieces to 23.62".

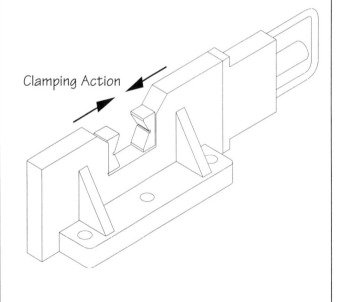

| | |
|---|---|
| *Description of Problem/Requirement:* | |
| Develop a universal workholding system for holding the widest range of workpiece variations. | |
| *Suggested Solution:* | |
| Incorporate a Gerardi Modular Vise into the workholding arrangement. | |
| *Source:* | |
| Koma Precision, Inc. | |

*Old Method:*

The plain machine vise, although very versatile, does lack many features necessary for holding some workpieces. Quite often, either the workpiece or the setup will require more than is possible with a standard machine vise. In these cases, either the vise must be modified, or custom fixturing is usually necessary.

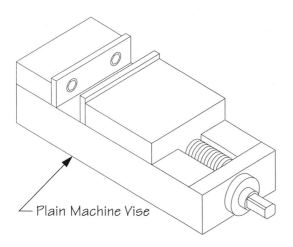

Plain Machine Vise

*New Method:*

The Modular Vise is a specialty vise well suited for many applications. The vise is available with several bases (4" to 12") and several jaw designs. The vise may be set up on a single base, or the jaws can be mounted on two bases for holding larger workpieces. The swivel base is also unique. Rather than a fixed-point swivel, the vise will also slide within the base for very precise positioning. A positioning pin can also be set in the base to allow the vise angle to be set with gauge blocks. The jaw elements are designed to pull the workpiece down as the vise is tightened, to assure the workpiece is firmly seated on the base.

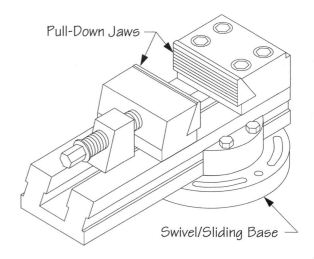

Pull-Down Jaws

Swivel/Sliding Base

# CHAPTER NINE

# Techniques for Power Workholding

# Techniques for Power Workholding

**Description of Problem/Requirement:**
Devise alternative workholding methods to enhance the capabilities of the operator.

**Suggested Solution:**
Incorporate power operated devices into the workholder designs.

**Source:**
Various jig and fixture component manufacturers.

**Old Method:**

In jig and fixture work, the methods and devices used for locating and holding workpieces are usually divided into two general categories: manual workholding systems and power workholding systems. Many of the more common elements used in constructing workholders can be used for either manual or power applications. Likewise, many of the devices employed in power workholding systems are simple modifications or variations of similar devices used for manual systems.

In the broadest sense, the main difference between the two systems is in the way the devices are operated. Manual workholding components are operated manually. The force directed to the workpiece in a manual system is determined by the force applied by the operator. Power workholding systems, on the other hand, although often initiated manually, apply force to the workpiece through a variety of power sources.

Manual workholding systems use a variety of mechanical mechanisms to apply force to the workpiece. Typically these include screw threads, wedges, and cams. These clamping devices rely on the mechanical advantage of these mechanisms as well as leverage to apply the forces necessary to securely hold the workpiece.

**New Method:**

Power workholding is a general term that describes a variety of methods and techniques used for positioning and clamping workpieces. Power workholding systems and devices have long been a part of manufacturing. However, not until newer self-contained power workholding systems became readily available has power workholding begun to emerge as a practical and viable workholding alternative.

Power workholding systems rely on a variety of sources to derive the force necessary to position and hold workpieces. The five principle types of power sources used for workholding applications are:

- hydraulic devices
- pneumatic (air) devices
- air-assisted hydraulic devices
- magnetic devices
- vacuum devices

Each of these is well suited for a variety of workholding tasks. Although each has applications where they are best suited, the workpiece and the application will usually determine the type of device used for any workholder.

*Description of Problem/Requirement:*

Develop a standardized method of holding workpieces for machining operations.

*Suggested Solution:*

Design the workholding methods around standard air-assisted hydraulic workholding systems.

*Source:*

Various jig and fixture component manufacturers.

*Old Method:*

The term power workholding covers a wide range of different workholding methods and systems. However, today, when the term power workholding is used, it most often describes either pneumatic (air) devices or air-assisted hydraulic workholding systems. Straight hydraulic workholding systems are very seldom used today. Likewise, when the other forms of power workholding systems—magnetic devices and vacuum devices—are used, they are referred to by their specific names.

The older, straight hydraulic workholding systems required the hydraulic fluid to be piped throughout the shop at very high pressures. This was not only messy but dangerous as well. The newer, air-assisted hydraulic system and the straight pneumatic (air) clamping system both use the standard shop air systems as their power sources.

*New Method:*

The air-assisted hydraulic systems most commonly used for workholding can be divided into three groups of elements. The shop air elements include the air inlet, a filter/regulator/lubricator device, safety valve, and a clamp/release valve. The shop air system typically has pressures of 90 to 100 psi (pounds per square inch). The next group of elements is the hydraulic system. This consists of a booster or pump, check valve (or other type of valve), and the manifold. This area is where the shop air is hydraulically boosted to approximately 3000 psi to operate the hydraulic components. The last group is the clamping system; it includes the various elements used to hold, position, and support the workpiece.

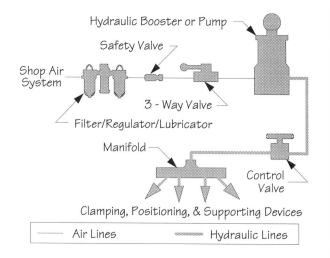

# Techniques for Power Workholding

**Description of Problem/Requirement:**
Develop a simplified standard system of preparing design drawings for power workholding systems.

**Suggested Solution:**
Incorporate a standard system of schematic symbols to replace drawn details in design drawings.

**Source:**
Various jig and fixture component manufacturers.

**Old Method:**

When designing either a pneumatic or air-assisted hydraulic power workholding system, the designer should prepare two different design drawings. The first drawing is a standard engineering drawing showing the complete workholder. This drawing contains the baseplate, locators, supports, and clamps. Each of these elements is drawn in its proper position in relation to the workpiece.

The second form of drawing that should be prepared for these workholders is a separate drawing of the power workholding system. The individual power components should only be drawn as exact details if a CAD database of the power components is available. Here, it is usually a simple matter to add each component to the drawing. But if each component must be drawn individually, the time required to prepare the power workholding system drawing can become excessive.

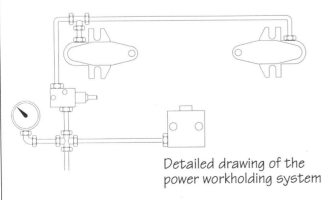

Detailed drawing of the power workholding system

**New Method:**

A better way to show the specific information about the power workholding system is with a circuit schematic drawing. This type of drawing uses a series of standardized symbols to show the various components. These circuit schematic drawings show the complete system and how each element is related to the others. This drawing is necessary to assure that each of the components specified is properly identified. A circuit schematic also reduces errors and allows the designer to spot any problem areas before the workholder is constructed. These drawings are also much simpler and take less time to prepare than full component drawings.

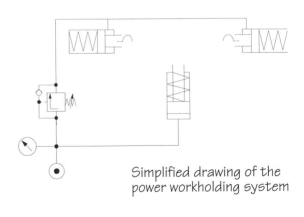

Simplified drawing of the power workholding system

*Description of Problem/Requirement:*

Design a plumbing system to increase the efficiency of the power workholding system.

*Suggested Solution:*

Consider using a manifold mounting style of plumbing system.

*Source:*

Various jig and fixture component manufacturers.

*Old Method:*

When constructing any workholders using an air-assisted hydraulic system, each component is operated with hydraulic fluid. The method used to supply hydraulic fluid to the components is an important part of the complete design.

Tubing on top of the fixture base

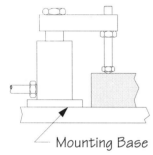

Mounting Base

Tubing below the fixture base

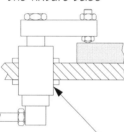

Threaded Body

The most common method of plumbing these fixtures is with the tubing on top of the fixture base. This method requires less machining of the fixture base, and reduces the building time. But chips are easily trapped around the lines and the exposed tubing can be damaged. The tubing may also be placed under the fixture base. This reduces any possible damage and prevents chips from building up in the work area. However, larger fixture bases are required, and the construction is more complicated.

*New Method:*

Manifold mounting is an alternative method of plumbing fixtures that does not require tubing. Here, the hydraulic fluid is supplied to the components via a series of holes through the fixture body. There are two styles of manifold mounting used for workholding applications.

"O"-Ring-Type Manifold Mount

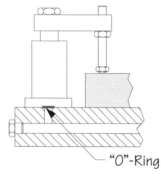

"O"-Ring

Cartridge-Type Manifold Mount

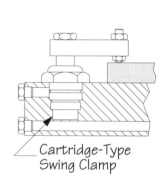

Cartridge-Type Swing Clamp

The first type uses passages drilled in the fixture body to feed fluid directly to "O"-ring ports under the clamps. The second style has a similar design, except that cartridge-type clamps are used. These clamps are embedded in special mounting holes. Both styles of manifold mounts prevent chip traps and create more compact workholders. Likewise, the fixtures tend to be more rigid since thicker bases are needed. However, this design does limit the positions of the clamps, and the passages must sometimes be gun drilled. This can add to the cost of the workholder.

# Techniques for Power Workholding

**Description of Problem/Requirement:**

Design a simpler plumbing system that is faster and easier to install.

**Suggested Solution:**

Replace solid tubing with flexible hydraulic hose to simplify the plumbing system.

**Source:**

Various jig and fixture component manufacturers.

**Old Method:**

The specific type of plumbing used for any workholder is often decided upon by the type of fixture and the projected workholding application. All four of the standard plumbing styles work well, and each is suitable for a wide range of standard and special workholders. The drilled holes for the manifold mounts obviously take more time to prepare than simply attaching the solid tubing. However, even the solid tubing takes time to cut, bend, and fit to the workholder. For this reason, solid tubing is best applied to workholders made for more permanent or long-term applications.

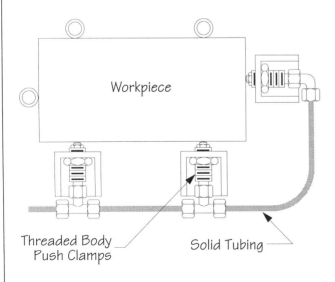

**New Method:**

To reduce the time and subsequent cost of plumbing an air-assisted hydraulic clamping system, flexible hydraulic hose should be used in place of solid tubing where practical. Depending on the manufacturer, flexible hose is readily available in several standard lengths from 9" to 15'. Special lengths are also available for any nonstandard applications. Flexible hoses should be used in place of solid tubing for short run applications, or for workholders that are not expected to be permanent. These hoses take considerably less time to install since they do not require cutting or fitting to specific lengths. Likewise, since they are flexible, they do not need to be bent to suit any installation.

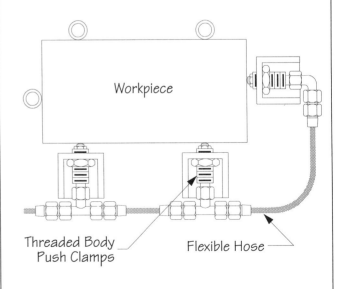

*Description of Problem/Requirement:*

Design a faster method to change the components in a power workholding system.

*Suggested Solution:*

Replace the threaded plumbing connections with quick disconnect-type fittings.

*Source:*

Various jig and fixture component manufacturers.

*Old Method:*

The fittings typically used for the plumbing connections in most power workholding systems use threaded connections. Depending on the manufacturer and the style of component, these connections may have different styles of threaded connections. The most common types of connections are those with flair-type fittings; however, pipe threads are also used. Regardless of the style of fitting, the major drawback to using standard fittings is in the time required to change the workholders. Also, these fittings typically require wrenches to connect or disconnect the hoses or tubing.

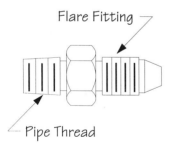

*New Method:*

A faster method of making the changeover between workholders is with quick disconnect couplers. These fittings are made in two styles: internal and external. The internal fitting is most often mounted to the workholder, while the external fitting is attached to the hose. Likewise, there are two types of quick disconnect fittings used for many workholders. One type is intended for air connections, while the other is used for hydraulic connections.

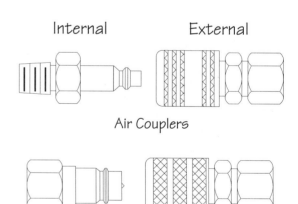

# Techniques for Power Workholding

**Description of Problem/Requirement:**

Devise a universal mounting system for the power workholding components.

**Suggested Solution:**

Design a grid plate and clamp adapter system for mounting the components.

**Source:**

Various jig and fixture component manufacturers and in-house fabrication.

**Old Method:**

Surface mounting is one of the more common methods used to attach power clamps to the fixture base. Here, the mounting holes for the power clamps are machined in the fixture base to suit the specific locational requirements of the workpiece. The clamps are then secured to the base with mechanical fasteners. This method, although very efficient, is best suited for permanent workholders that are dedicated to a single workpiece.

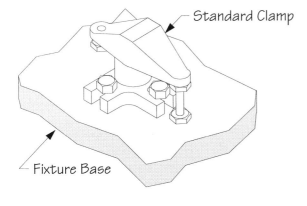

**New Method:**

An alternate method of attaching power clamps is with an adapter plate and grid plate arrangement. Here, the clamps are mounted on an adapter made to suit both the clamp mount and the hole pattern of the grid plate. Slots in the adapter allow positional adjustment for the clamps. This arrangement allows a wide range of potential workholding setups. Several clamps may be mounted to the grid plate to hold the workpieces. Quick disconnect fittings should also be used for the clamps and the supply manifold to speed the clamp changeover.

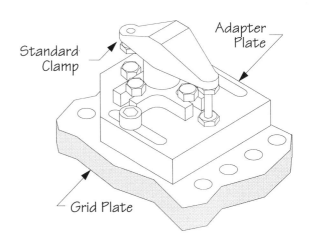

| Description of Problem/Requirement: |
|---|
| Design an efficient and cost-effective workholding system for mounting multiple workpieces. |

| Suggested Solution: |
|---|
| Design the workholder around the Flexible Clamping System. |

| Source: |
|---|
| Carr Lane Manufacturing Co. |

**Old Method:**

Mounting multiple workpieces can greatly increase the overall efficiency of any production setup. However, unless the proper methods are used for fixturing these workpieces, the gains made from machining multiple workpieces can easily be offset with excessive changeover times. Threaded fasteners are one way these workpieces may be clamped, but the time required to clamp and unclamp each workpiece can be considerable when several workpieces are loaded in the workholder.

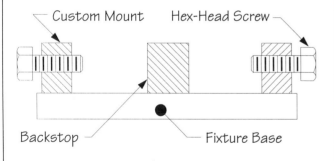

**New Method:**

An alternative method of fixturing these multiple workpieces is with the Flexible Clamping System. This system uses a standard size baseplate and a series of manifold mounted threaded body push clamps to hold the workpieces. These elements are completely adjustable and may be positioned where needed on the baseplate. Since the clamps in this unit are mounted in a manifold, all the clamps are operated from a single point. So, when clamped, all the clamps operate at the same time. Likewise, when unclamped, all the clamps open at the same time.

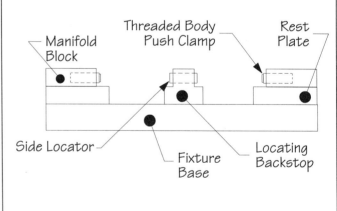

# Techniques for Power Workholding

**Description of Problem/Requirement:**

Design an efficient and cost-effective workholding system for mounting multiple workpieces.

**Suggested Solution:**

Design the workholder around the Flexible Clamping System.

**Source:**

Carr Lane Manufacturing Co.

**Old Method:**

When fixturing multiple workpieces, the specific number of workpieces fixtured is most often determined by the size of the baseplate and the sizes of the workpieces. Larger baseplates will typically fixture more workpieces. However, when large workpieces are fixtured, the increased size will often limit the number of units that can be mounted in any single setup. With custom made workholders, almost any number of workpieces may be set up by simply changing the size of the baseplate. But, as with any custom workholder, the cost may exceed the benefit derived from using the workholder.

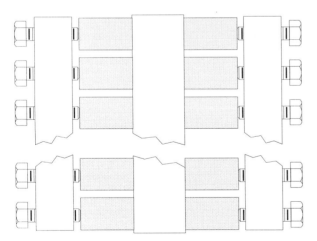

Number of workpieces in any setup is usually decided by the baseplate and workpiece sizes.

**New Method:**

The Flexible Clamping System uses a standard baseplate and, as such, has a limited range of possible setups. Although quite adaptable, this fixed size baseplate often dictates the number of workpieces fixtured in any single setup. While several smaller workpieces may be fixtured, fewer larger workpieces may be mounted in any single setup. However, despite this, the cost savings derived from using this system will often offset the occasional production restrictions.

Several smaller workpieces in any setup:

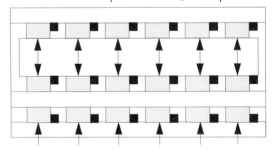

Fewer larger workpieces in any setup:

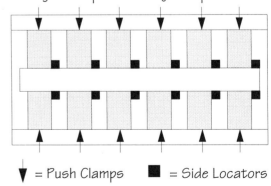

▼ = Push Clamps   ■ = Side Locators

*Description of Problem/Requirement:*

Devise an efficient method of mounting multiple workpieces in specialized workholders.

*Suggested Solution:*

Design the workholder around manifold mounted, threaded body push clamps.

*Source:*

Various jig and fixture component manufacturers and in-house fabrication.

*Old Method:*

Threaded body push clamps are one of the more efficient and useful types of hydraulic clamps. Their small size and overall design permit them to be positioned almost anywhere on a workholder. Standard mounting brackets are available for attaching these clamps to the fixture base. These brackets will reduce the time required to construct the workholder, and are very well suited for workholders having just a few clamps. When several clamps are needed, more time is required to run the tubing or hose between the clamps.

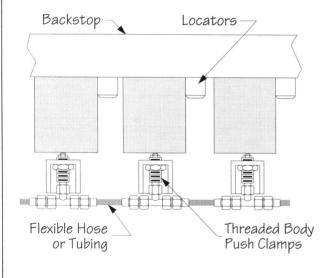

*New Method:*

An alternative method of plumbing these threaded body push clamps is with a manifold block. Here the clamps are threaded into a block at the desired locations. Rather than using tubing or hoses to supply the hydraulic fluid, a single hole is drilled through the block to interconnect with the cylinders. This arrangement is very well suited for specialized workholders. The manifold block can be made in almost any shape and permits the clamps to be positioned closer together. This also eliminates the mounting brackets and plumbing each clamp with tubing or hose.

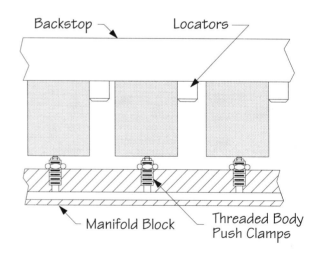

# Techniques for Power Workholding

**Description of Problem/Requirement:**

Develop a method of retrofitting existing manual strap clamps to suit power clamping systems.

**Suggested Solution:**

Incorporate a thru-hole cylinder-type hydraulic clamp into the workholder design.

**Source:**

Various jig and fixture component manufacturers.

**Old Method:**

Strap clamps are some of the most common clamping devices used for workholders. The basic design of the strap clamp directs the clamping force of a threaded fastener (or a cam) through a flat clamping bar to the workpiece. The clamping action is accomplished at one end of the bar, while a heel pin or similar device at the opposite end acts as a fulcrum. Due to their simplicity and lower cost, strap clamps are the most popular type of clamping devices used for workholding.

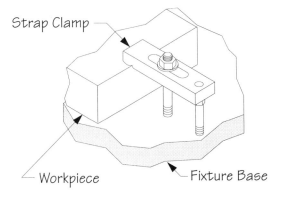

**New Method:**

To increase production rates or volume, it is occasionally desirable to convert existing manual workholders to power workholding devices. Fixtures that use strap clamps are relatively easy to retrofit by simply adding a thru-hole cylinder-type power clamp. Here the thru-hole cylinder clamp is mounted over the clamping stud in the existing strap clamp arrangement. Depending on the specific clamp design, the stud may either be threaded into the clamp plunger or inserted through the hole in the hollow plunger and held with a nut. A longer clamping stud is often required when a thru-hole cylinder clamp is installed.

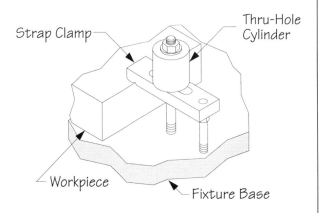

**Description of Problem/Requirement:**

Devise simplified methods of clamping workpieces with power clamping devices.

**Suggested Solution:**

Design the workholders to use thru-hole cylinders to clamp the workpieces.

**Source:**

Various jig and fixture component manufacturers.

**Old Method:**

Power clamping offers a variety of benefits and advantages. With just a little imagination and forethought, power clamps may be applied in a variety of creative setups. The thru-hole cylinder-type clamp is one of the more versatile and functional styles of power clamps.

The workpieces shown here illustrate how the thru-hole cylinders may be applied to specialized clamping situations. Although any of several styles of power clamps may be employed to hold these workpieces, the thru-hole cylinder may be used for both, by simply modifying the ways they are applied.

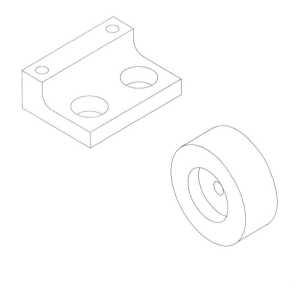

**New Method:**

This workholder must be designed to minimize any obstructions above the workpiece. Here the thru-hole cylinders are mounted below the fixture base. In use, they pull the clamping studs down to clamp the workpiece.

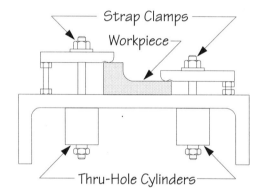

This workholder is designed to clamp the workpiece against the vertical face of the angle. The clamp is installed in the center hole and pulls the workpiece back when the pressure is applied. A "C" washer is used to permit easier loading and unloading of the workpiece.

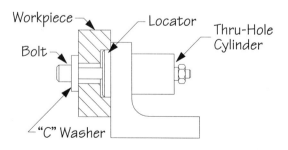

# Techniques for Power Workholding

*Description of Problem/Requirement:*

Design a supporting system to precisely and securely position the workpiece.

*Suggested Solution:*

Incorporate a combination of mechanical and power work supports into the workholder design.

*Source:*

Various jig and fixture component manufacturers and in-house fabrication.

*Old Method:*

Supporting the workpiece is a primary function of any workholder. Two of the more common work supports used for many workholders are solid supports and manually adjustable supports.

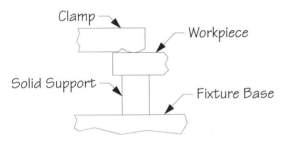

Solid supports are usually simple blocks, machined to the required height for the workpiece feature. When workpieces are mounted directly to a fixture base, the baseplate itself is a solid support.

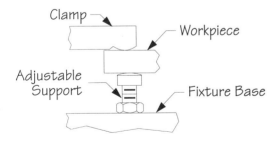

Adjustable supports are often used for cast or forged workpieces where a degree of adjustment to suit each workpiece is desired. These supports may also be used for fixed height support. Here they are simply adjusted to the desired height and locked in place.

*New Method:*

Rather than relying on mechanical supports alone, a better choice for many workholders is a combination of mechanical and power work supports. Here, the mechanical supports can fix the location of the workpiece, while the power supports can be used to support the workpiece at specific points to prevent distortion during the machining operations.

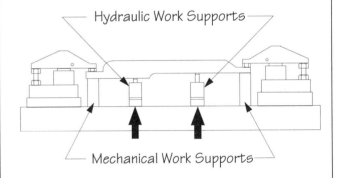

The three primary forms, or actions, of work supports are the spring extended, fluid advanced, and air advanced types. Each of these is available in several shapes and mounting configurations to meet their many applications. Each of these work supports automatically adjusts to the correct workpiece height and, once positioned, is securely locked under hydraulic pressure. In many clamping arrangements, a sequence valve is used to first position the supports and then to lock the clamps.

**Description of Problem/Requirement:**

Devise a method of using power workholding systems with pallet setups.

**Suggested Solution:**

Incorporate a pallet decoupler into the workholder design.

**Source:**

Various jig and fixture component manufacturers.

**Old Method:**

Pallet setups are becoming quite popular in manufacturing today. In fact, most newer machine tools are available with a variety of different pallet arrangements. However, when using normal power clamping systems, devising a method to use both pallets and power clamping can be tricky.

This is due to the fact that the power workholding system must normally remain connected to its power source to maintain the hydraulic pressure. The one unalterable rule of power clamping is: if there is no hydraulic pressure, there is no clamping pressure. This is also true for machine tools that have doors that contain the workpiece during the machining cycle. Many times, in these situations, manual clamping systems were used simply because there was no way to maintain the hydraulic connections.

**New Method:**

Pallet decouplers allow the workholder to remain pressurized when disconnected from the power source. These decouplers have an accumulator built into the unit. Accumulators use either fluid or gas to maintain the pressure in the system, when disconnected from the power source. In use, the system is pressurized and the shutoff valve is closed. The supply hose is then removed. The accumulator retains the required pressure in the system. A pressure gauge on the unit shows the exact pressure in the system. There is a variety of manual and automatic pallet coupling devices commercially available.

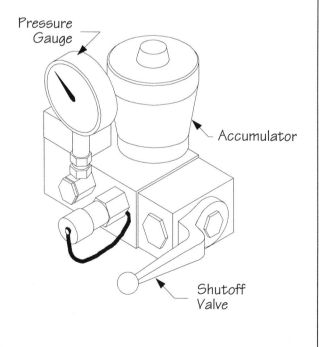

# Techniques for Power Workholding

**Description of Problem/Requirement:**
Design a method of providing a positive mechanical lock for a power workholding system.

**Suggested Solution:**
Design the workholder with StayLock™ clamps to hold the workpiece.

**Source:**
Jergens Incorporated.

**Old Method:**

Pallet decouplers and other similar applications of accumulators work well for many workholding applications. But, accumulators, by design, have one distinct limitation. These units are only intended to maintain the existing system pressure. If anything happens to the system, such as a leak or a broken tube or hose, where the hydraulic fluid is lost, the accumulator cannot maintain the system pressure. If there is no hydraulic pressure, there is no clamping pressure.

Aside from the fact that this condition can cause delays and create other problems during machining, the safety aspects of these setups are also a prime consideration. From strictly a safety point of view, the clamps must remain engaged no matter what happens to the power supply or hydraulic fluid. Unfortunately, all power workholding systems do have this inherent design limitation—to operate, they must remain pressurized.

**New Method:**

One answer to the problem of pressurization in power workholding is the StayLock™ clamp. This clamp design uses a unique mechanical locking arrangement to lock the clamps. These clamps are hydraulically activated, mechanically locked, and hydraulically deactivated. Once the locking angle is engaged on the locking pin, the system no longer relies on the hydraulic pressure. In fact, the hoses may be removed, and all the clamps will remain locked until they are hydraulically unlocked. This StayLock™ principle is available in a variety of clamp styles.

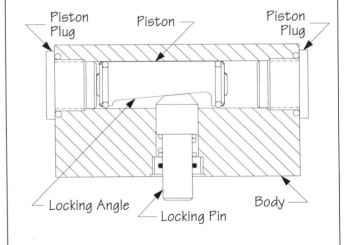

**Description of Problem/Requirement:**

Devise an alternative power workholding system for use where hydraulic clamps may be objectionable.

**Suggested Solution:**

Design the workholder using a pneumatic workholding system.

**Source:**

Various jig and fixture component manufacturers.

**Old Method:**

Another form of power workholding system that is widely used throughout manufacturing today is pneumatic, or air-operated, workholding systems. Unlike air-assisted hydraulic systems, these clamps and accessories operate completely on air pressure.

Pneumatic clamping systems are well suited for applications where hydraulic fluid could cause contamination, or other problems. There are many variations of pneumatic clamps. These clamps are very well suited for applications where space is a problem. Typically, pneumatic clamps are used for light duty machining, inspection setups, assembly work, and similar operations.

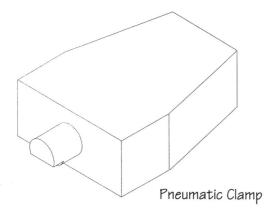

Pneumatic Clamp

**New Method:**

Although pneumatic workholding systems are gaining wider acceptance, the primary drawback to their widespread application in heavy machining is the size of the ram needed to generate the holding forces. The surface area of the typical pneumatic cylinder must be approximately five times larger than a hydraulic cylinder to generate the same force. So, if a hydraulic ram has a surface area of 1 square inch, to generate the same force, the pneumatic ram will require 5 square inches of surface area.

Another problem with pneumatic clamps is in the air itself. Unlike hydraulic fluid that will not compress under a load, air used for pneumatic clamps will compress. For applications where the compression of the air could cause workholding problems, hydraulic systems are generally preferred.

# Techniques for Power Workholding

**Description of Problem/Requirement:**

Devise an improved method of holding ferrous workpieces with a magnetic chuck.

**Suggested Solution:**

Use an electromagnet magnetic chuck rather than a permanent magnet magnetic chuck.

**Source:**

Various jig and fixture component manufacturers.

**Old Method:**

Magnetic chucks are commonly used for a variety of workholding applications where the machining stresses are light to moderate. Surface grinding is one operation where the magnetic chuck is the most common form of workholding device. But magnetic chucks can also be used for heavy duty machining applications. Here, the only requirement is that the correct chuck must be used.

The most common form of magnetic chuck is the permanent magnetic type. These chucks are readily identified by the swing handle used to turn the chuck on and off. These chucks, although well suited for light work, cannot generate enough holding force for heavier machining operations.

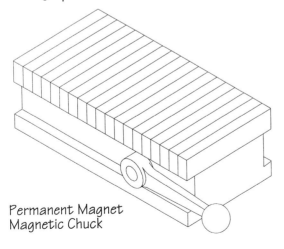

Permanent Magnet Magnetic Chuck

**New Method:**

For heavier machining operations, an electromagnet magnetic chuck is the better choice. These chucks use electricity to generate their holding force. Electromagnet magnetic chucks can be identified by an electrical wire rather than a handle. Likewise, these chucks are turned on and off with an electrical switch. Although less commonly found in machine shops, these chucks offer considerably more holding force than the permanent magnet-type chucks.

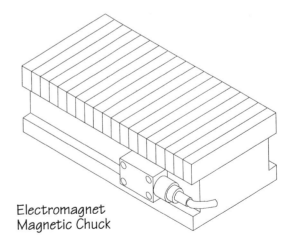

Electromagnet Magnetic Chuck

| Description of Problem/Requirement: |
|---|
| Design a simplified and secure system for holding ferrous workpieces on magnetic chucks. |

| Suggested Solution: |
|---|
| Incorporate a system of packing around the workpiece. |

| Source: |
|---|
| In-house fabrication. |

| Old Method: | New Method: |
|---|---|
| There is a variety of methods used for mounting workpieces on a magnetic chuck. Some workpieces may be mounted in a fixture, vise, or similar mechanical workholder, mounted on the magnetic chuck. But the most common mounting method is to simply position the workpiece directly on the face of the chuck and engage the magnet.<br><br>Regardless of the mounting methods, it is important to remember that a magnetic chuck only holds in one direction—downward. The workpiece itself must have sufficient mass to resist any side thrusts. | For workpieces without sufficient mass (or size) to resist the side forces, a series of additional elements may be positioned around the workpiece. These elements are called packing and help the workpiece resist any side thrust forces encountered in the machining operation. The packing may be made from any ferrous material and may be any size. The only requirement is that the height of the packing should be less than the height of the workpiece. This will help isolate the workpiece and prevent the packing from interfering with the machining operations. |

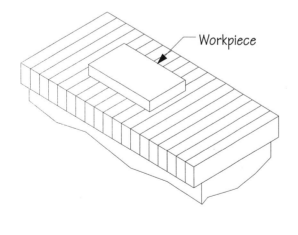

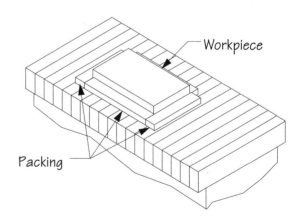

# Techniques for Power Workholding

**Description of Problem/Requirement:**

Design a simplified and secure system for holding nonferrous workpieces on magnetic chucks.

**Suggested Solution:**

Incorporate a system of custom clamps as packing around the workpiece.

**Source:**

In-house fabrication.

**Old Method:**

Magnetic chucks are typically used for holding ferrous workpieces. However, there are times when nonferrous workpieces must also be mounted on magnetic chucks. The magnetic forces generated by a magnetic chuck have no effect on nonferrous workpieces. Here, some type of auxiliary clamping device must be mounted to the magnetic chuck and used to hold the workpiece.

Plain packing placed around the periphery of the workpiece will sometimes not provide sufficient support for the workpiece during the machining operation. So, another type of arrangement must be employed to secure the workpiece.

**New Method:**

For those applications where greater machining forces are expected, plain packing may not provide sufficient support to properly restrain the workpiece.

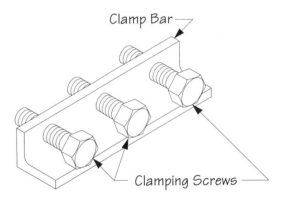

Here, a custom clamping bar arrangement may be a better alternative. These bars may be made in a variety of sizes and forms to accommodate a variety of nonferrous workpieces. Regardless of the basic form of the unit, the base of the bar is designed to be clamped by the magnetic chuck, while the screws are intended to clamp the workpiece.

# Techniques for Power Workholding

| Description of Problem/Requirement: |
|---|
| Design a simplified vacuum-type workholding system for holding nonferrous workpieces. |

| Suggested Solution: |
|---|
| Incorporate a Dunham vacuum chuck into the workholding system in place of a standard-type chuck. |

| Source: |
|---|
| Dunham Tool Company. |

### Old Method:

Magnetic chucks, although well suited for many applications, are usually limited to holding only ferrous workpieces or accessories. For those situations where ferrous workpieces must be held, another type of chuck may be used. This chuck is the vacuum chuck.

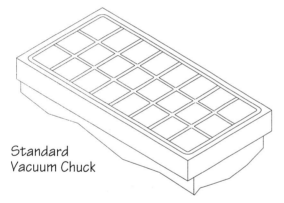

Standard Vacuum Chuck

Vacuum chucks are designed to hold the workpiece against the face of the chuck by removing all the air from between the workpiece and the chuck, thus forming a vacuum. This vacuum can generate between 12 and 14 pounds per square inch of holding force across the face of the chuck. Standard vacuum chucks are designed so the entire face of the chuck engages the workpiece. In those cases where the workpiece is smaller than the complete chuck, a mask must be made to cover the areas of the chuck face not in contact with the workpiece. These masks help establish the vacuum and prevent the chuck from drawing in outside air.

### New Method:

Another style of vacuum chuck well suited for a variety of clamping operations on smaller workpieces is the Dunham vacuum chuck. Unlike the standard chuck design that uses grooves in the face of the chuck to draw the vacuum, this chuck uses a series of holes.

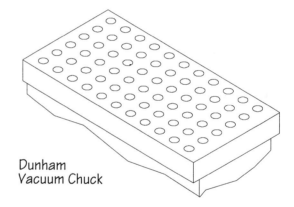

Dunham Vacuum Chuck

These holes are all threaded and fitted with special self-sealing screws. When the screws are loosened, the air freely flows through the screw to create the vacuum. However, when tightened, the hole is sealed and no air can pass through the screw. This design eliminates the need for a mask and allows the vacuum force to be directed to any area on the chuck face by simply opening only those screws under the workpiece.

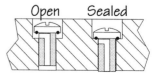

Open   Sealed

# Techniques for Power Workholding

**Description of Problem/Requirement:**

Design a method of holding irregular workpieces with a vacuum chuck.

**Suggested Solution:**

Incorporate an epoxy mounting block cast to suit the form of the workpiece.

**Source:**

In-house fabrication.

**Old Method:**

Standard vacuum chucks, like magnetic chucks, are made with a flat clamping face. Although appropriate for most workpieces, there are some workpieces that cannot be mounted on any workholder with a flat mounting surface. Some of the more common problem workpieces are nonferrous castings and forgings. Here, the shape of the workpiece cannot be securely held on the flat face of the vacuum chuck.

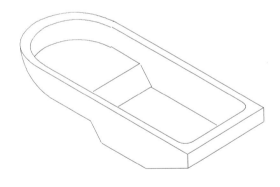

**New Method:**

To hold these odd-shaped workpieces, another device can be added to the vacuum chuck. This is a simple cast epoxy block. These blocks are first cast around the required workpiece shape. Once the epoxy is set, the block is machined to add the "O"-ring groove and the vacuum groove. Holes are also drilled through the block to both mount the block to the vacuum chuck and to permit the vacuum to be drawn through the block. The workpiece is mounted in the block and depressed against the "O"-ring. The vacuum pump is then engaged to draw the vacuum that holds the workpiece against the block.

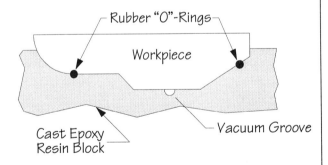

# PART THREE
## Additional Resources

# Company Names and Addresses

# Company Names and Addresses

Carr Lane Manufacturing Co.
4200 Carr Lane Court
St. Louis, MO 63119


Chick Machine Tool, Inc.
800 Commonwealth Drive
Warrendale, PA 15086


Clamp Chuck Systems
P.O. Box 219
Tiller, OR 97484


Dapra Corporation
66 Granby Street
Bloomfield, CT 06002


Discount Tool
12100 West 52nd Ave., Unit 123
Wheat Ridge, CO 80033
303-420-4007


Dunham Tool company
Drawer D
New Fairfield, CT 06812
203-746-2407


EDI Tooling Systems
5029 Willow Creek Road
Machesney Park, IL 61115


Enterprise Prohold Company
3316 Montgomery
Cincinnati, OH 45207


Hardinge Brothers, Inc.
Elmira, NY 14902


Hilma Corporation
(see Carr Lane
Manufacturing Company)


Huron Machine Products
228 Southeast 21st Terrace
Ft. Lauderdale, FL 33312


Interlen Products Corporation
1106 Second Street, Suite 811
Encinitas, CA 92024


Jergens Incorporated
19520 Nottingham Road
Cleveland, OH 44110


Koma Precision, Inc.
P.O. Box 628
Warehouse Point, CT 06088

Mitee-Bite Products Inc.
P.O. Box 477
Old Route 16
Center Ossipee, NH 03814

Morgan Enterprises
P.O. Box 299
Waldron, IN 46182

James Morton Inc.
P.O. Box 399
Batavia, NY 14021

Powerhold, Inc.
P.O. Box 447
Middlefield, CT 06455

ROVI Products
4685 Runway St., Unit G
Simi Valley, CA 93603

Royal Products
210 Oser Avenue
Hauppauge, NY 11788

Safe-Tech Corporation
729 Thomas Drive
Bensenville, IL 60106
708-238-0715

SMW Systems Inc.
9828 South Arlee Ave.
Santa Fe Springs, CA 90670

Stevens Engineering Inc.
3946 W. Clarendon Ave.
Phoenix, AZ 85019

Tayco Tools, Inc.
P.O. Box 412
Michigan Center, MI 49254

Turnmill Machine Company
33215 Dequindre
Troy, MI 48083

S.B. Whistler & Sons
P.O. Box D
Buffalo, NY 14217

# Index

**A**
Accumulators, 224, 225
Adhesive Clamping, 114
Adjustable Work Stop, 190
Alignment Pin, 133–135, 184

**B**
Ball Lock Mounting System, 141–143
Ball Plunger Assembly, 74
Baseplates, 102, 122, 123, 142, 206, 218, 219
Boss on Castings/Forgings, 11

**C**
Casting, 10–12
    Boss, 11
    Tabs, 11, 94
Chamfer, 10, 15, 16
Chuck, Magnetic, 227–231
Chuck, Magnetic Electromagnetic, 227
Chuck, Magnetic Permanent Magnet, 227
Chuck, Step, 166, 167, 170, 171, 174
Chuck, Vacuum, 230
Chuck Jaws, 147–149
Chuck Jaws, Master Jaw System, 161
Chuck Jaws, Quick Change, 152, 153
Chuck Jaws, Universal Forming Device, 150
Chuck Jaws, Universal Turning Fixture, 151
Chuck Mounting Systems
    Clamp Chuck System, 155–157
    Prohold System, 154
Clamp Function 19, 21
Clamping, Adhesive, 114
Clamping Cylindrical Workpieces, 188, 200, 207
Clamping Distortion, 26, 91, 94, 185, 223
Clamping Forces, 26, 87, 188, 189, 199
Clamping Pallet Arrangements, 97–99
Clamping Positions, 26, 27
Clamps, 19, 26, 27
    Ball Element, 91, 94

Cam, 27, 28, 105, 106, 107
Claw, 116
Edge, 95, 100, 107, 109, 115
Edge, Flat, 96
Edge, Pivoting, 95
Edge, Slot, 96
Edge, Two Directional, 102
Epoxy/Subplate, 114
Flexible Clamping System, 218, 219
Grip Strip System, 115
Hi-rise, 90
Hook, 92
Hydraulic, Thru-Hole Cylinder, 221, 222
    and magnetic chucks, 229
Mono Bloc, Fast Acting, 103, 104
Multiple Workpiece, 97–99, 114, 115, 189, 191, 196–198, 218, 219
Pneumatic, 226
Screw, 27, 64, 92, 93, 96, 98, 101, 109
Screw, Quick Acting, 93
Spherical Washer Set, 86
Stay Lock, 225
Strap, 28, 35, 83–88, 90, 91, 103, 104, 126, 189, 198, 205, 207, 221, 222
Strap, Adjustable, 89
Swing, 92
Threaded Body Push, 218, 220
Toggle, 111, 112
Toggle, Automatic, 113
Up-thrust, 108
Wedge, 109, 110
Collet Accessories, 165, 168–171
Collet Stops, 163, 164
Collets
    Expanding, 158, 172–175
    5C Emergency, 158–160, 162, 164, 170, 171
    Master Jaw System, 159
    Step, 158

Cutter Setting Device, 192
Cutting Forces, 179, 181
Cylindrical Workpieces, 61, 62, 76, 183, 200, 207

**D**
Datums, 11–13, 23, 24
    Datum targets, 11
    Imaginary Datums, 12
    Physical Datums, 12
Dedicated Workholders, 36, 40–43, 139, 168, 175, 193, 195
Degrees of Freedom, 20–22, 75
    Axial, 20
    Radial, 20
Diamond Pin, 52–54
Dowel Pins, 49–51, 53, 131, 138, 141, 183, 187
Dowel Pins, Grooved, 131
Dowel Pins, Two-Piece, 132

**E**
Eccentric Leveling Lug, 128
Eccentric Liner, 71
Epoxy Mounting Block, 231

**F**
Fasteners, 130
Fixture Key, 119–121, 126
Flexible Clamping System, 218, 219
Foolproofing, 24
    Foolproofing Device, 24, 25, 77–79, 164
Force 14 Vise Accessory, 194
Forging
    Boss, 11
    Tabs, 11, 94

**H**
Hoist Rings, 137

**I**
Indexing, 73, 74
Indexing Devices
    5C Indexing Head, 160
    Indexing Plunger, 74
    Jig Pin, 73

**J**
Jig Pin, 73

**L**
Leveling Devices
    Eccentric Leveling Lug, 128
Locating, 19–23
    Cylindrical Workpieces, 61, 62, 76

    Duplicate, 23, 24
    Primary Feature, 19
    Primary Surface, 20, 21
    Principles, 19
    Redundant (See Locationg, Duplicate)
    Secondary Surface, 20, 21
    Six-Point Method (3-2-1), 20, 21
    Tertiary Surface, 20, 21
Locator Types
    Adjustable, 64
    Adjustable Ball Locking Pin, 135
    Alignment Pin, 133–135, 184
    Ball Lock Mounting System, 141–143
    Ball Plunger Assembly, 74
    Bushings, 136
    Conical, 58, 59
    Conical, Adjustable, 59
    Diamond Pin, 52–54
    Dowel Pins, 49–51, 53, 131, 138, 141, 183, 187
    Dowel Pins, Grooved, 131
    Dowel Pins, Two-Piece, 132
    Expanding Pin, 135
    Fixture Key, 119–121, 126
    Floating Locating Pin, 60
    Locating Pin, 56, 61, 79
    Locating Plug, 61
    Locating Ring, 61
    Locator/Bushing Arrangement, 51
    Lock Screw, 50
    Locking Pin, 134
    Modified Relieved, 56
    Nesting, 75, 76
    Nests, Dowel Pin, 76
    Nests, Epoxy Resin, 75
    Nests, Partial, 76
    Nests, Ring, 76
    Plain, 49
    Raised Contact, 55
    Relieved, 52
    Retractable Plunger, 73
    Right Angle, 125
    Roll Pin, 131
    Self-Centering, 58
    Shoulder, 49
    Single Axis, 126
    Solid, 59, 61, 64
    Spherical, 56, 57
    Spring Locating Pin, 70–72, 131
    "T" Nut/Dowel Pin Arrangement, 122–124
    "V"-Type, 62, 63, 182, 207
Locators, 19, 20, 22, 25, 26
    Concentric, 25, 60, 78, 79

# Index

Function, 19
Primary, 22
Radial, 22, 25, 54, 60, 78, 79
Low-Melt Alloys, 29, 75

## M

Machining Center Vise, 204
Magnetic Chuck, 227–231
Magnetic Chuck, Electromagnetic, 227
Magnetic Chuck, Permanent Magnet, 227
Manifold System, 214, 220
Master Jaw System, 161
Master Plate, 139
Maxi Mill Vise, 205
Maximum Material Condition (MMC), 22, 60
Modular Vise, 206
Modular Vise, Swivel, 208
Modular Workholders, 36–43
    Cost, 41–43
    Dowel Pin, 38, 39, 140
    Elements, 36–40, 139, 140, 206, 208
    Subplate Systems, 36, 37, 140–143
    T-Slot Systems, 36, 37, 140
Mono-Bloc Fast Acting Clamp, 103, 104
Multi-Fixture Vise, 200
Multi-Lok Vise, 202
Multiple Workpiece Setups, 97–99, 114, 154, 189, 191, 196–198, 201, 202, 218, 219

## N

Nest Types
    Dowel Pin, 76
    Epoxy Resin, 75
    Partial, 76
    Ring, 76
Nesting, 75, 76

## P

Pallet Arrangements/Systems, 97–99, 138, 139, 143
Pallet Decoupler, 224
Parallels, 187
Pneumatic Workholding System, 226
Power Workholding, 211
    Air Assisted Hydraulic Workholding, 212
    Construction, 213, 215, 217
    Flexible Clamping System, 218, 219
    Hydraulic Thru-Hole Cylinder Type, 221, 222
    Magnetic Chucks, 228–231
    Magnetic Chucks, Electromagnetic, 227
    Magnetic Chucks, Permanent Magnet, 227
    Manifold System, 214, 220
    Mechanical and Power Supports, 223
    for Pallet Systems, 224
    Pneumatic System, 226
    Quick Connect Fittings, 216
    Staylock Clamp, 225
    System Drawings, 213
    Vacuum Chuck, 230
    Vacuum Chuck used with Epoxy Mounting Blocks, 231
Product Designer, 7

## Q

Quad 1 Precision Vise, 199
Quick-Acting Screw Clamps, 93
Quick Change Chuck Jaws, 152, 153
Quick Change Fixtures, 143
Quick Connect Fittings, 217
Qwik Lok Vise, 201

## R

Radii, 15, 16
Retractable Plunger, 73

## S

Self-Centering Vise, 207
Setup Reduction
    Areas of Concern, 5
    Benefits, 5
    Overview, 5
    Principles, 35
Six-Point Method (3-2-1), 20, 21
Spotfacing, 10, 13, 14
Stay Lock Clamp, 225
Subplate Systems, 36, 37, 140–143
Support Types
    Adjustable, 66, 67, 223
    Buttons, 65, 72
    Elevating, 129
    Equalizing, 67
    Heel, 83, 85, 87, 89
    Jig Foot, 127, 129
    Mechanical and Power Combination, 223
    Screw, 227
    Spherical Contact, 68
    Spring Stop Button, 72
    Swivel Ball, 68
    Thrust, 69
Supports, 19, 22, 23, 27, 65, 66
Swivel Jaw Vise Accessory, 195

## T

"T" Nut/Dowel Pin Arrangement, 122–124
Tabs on Castings/Forgings, 11, 94
Temporary Workholders, 35, 36, 40

Toggle Clamps, 111–113
Tolerances, 9, 16, 22–31
    Arbitrary, 9, 22, 23
    Locational, 22
    Percentage, 9, 22, 23
    Product, 9
    Resulting, 9
    Tooling, 22, 23, 30, 31
    Workholder, 22, 23
    Workpiece, 23
Tool Bodies, 28–32
    Built-up, 29, 30
    Cast, 28, 29
    Built-up, 29, 30
    Designing, 30–32
    Materials, 30–32
    Welded, 29
Tool Forces, 26
    Primary, 26
    Secondary, 25–26
Tool Steel, 30–32
Tooling Holes, 12
Twin-Lock, 196–198

## U
Undercut, 10, 14

## V
Vacuum Chuck, 230, 231
Vacuum Chuck with Epoxy Mounting Blocks, 231
Vise, Forces, 181
Vise, Proper use, 181
Vise Accessories
    Adjustable Work Stop, 190
    Cutter Setting Device, 192
    Force 14 Vics Accessory, 184
    Parallels, 187
    Quick Change Machinable, 203
    Side Clamping Device, 189
    Swivel Jaw, 195
    Twin Lock, 196–198
Vise Jaws, Quick Change, 193
Vises, Specialty, 179
    Machining Center, 204
    Maxi Mill Vise, 205
    Modular Vise, 206
    Modular Vise, Swivel, 208
    Multi-Fixture Vise, 202
    Multi-Lok Vise, 202
    Quad 1 Precision Vise, 199
    Qwik Lok Vise, 201
    Self-Centering Vise, 207

## W
Workholder Types, 35
    Dedicated, 36, 40–43, 139, 168, 175, 193, 195
    Modular, 36–43, 139, 140
    Temporary, 35, 36, 40
Workholding
    Costs, 41, 43
    Design Guidelines, 40
    Options, 35
    Principles, 19–32
    Repeatability, 19
        in Setup Reduction, 5, 7, 8
    Selecting, 35, 39–41, 43
Workholding, Power, 211
    Air Assisted Hydraulic Workholding, 212
    Construction, 213–215, 217
    Flexible Clamping System, 218, 219
    Hydraulic Thru-Hole Cylinder Type, 221, 222
    Magnetic Chucks, 228–231
    Magnetic Chucks, Electromagnetic, 227
    Magnetic Chucks, Permanent Magnet, 227
    Manifold System, 214, 220
    Mechanical and Power Supports, 223
        for Pallet Systems, 224
    Pneumatic System, 226
    Quick Connect Fittings, 216
    Staylock Clamp, 225
    System Drawings, 213
    Vacuum Chuck, 230
Workpiece Design, 8, 10
Workpiece Materials, 8–9
Workpiece Processing, 13–14

## X
X-axis, 19, 20, 59, 60

## Y
Y-axis, 19, 20, 59, 60

## Z
Z-axis, 19, 20, 59